1981

$8.95

JACK DENTON SCOTT

WINDOW ON THE WILD

with linocuts by
Geri Greinke

G. P. PUTNAM'S SONS NEW YORK

Printed in the United States of America
First Impression
Design by Kathleen Westray
Library of Congress Cataloging in Publication Data
Scott, Jack Denton.
Window on the wild.
Summary: Discusses 11 animals that the author has personally observed: the earthworm, monarch butterfly, beaver, badger, red fox, gray squirrel, weasel, ruffed grouse, bald eagle, loon, and great horned owl.
1. Animals—Miscellanea—Juvenile literature.
2. Wildlife watching—Juvenile literature. [1. Animals. 2. Wildlife watching] I. Greinke, Geri.
II. Title.
QL49.S29 1980 591 80-15250
ISBN 0-399-20722-8

For

MARIA LUISA,

who climbs my mountains
and shares my tent—
always with pencil handy
and observations keen,
which makes this as
much her book as mine.

Contents

I HAD the good fortune to be taught to see in the wild places by professionals in India and Africa, who, in turn, had received lessons from the best watchers in existence, the native people of those wild regions.

Today, finally, after we have placed in danger of extinction more than 1,000 breeds of wild creatures worldwide, completely wiped out 50 species, and, in the United States alone, seriously endangered 109 wild animals and birds, appreciative watching—not hunting or harming—is on the rise everywhere.

There are no legal bag limits on taking with the eyes as many grouse as you can find flying through the woodland, none on capturing with your own wide-ranging vision an agile tree-jumping squirrel, or, if you are lucky, snap-shooting with the lens of your camera or eye even the bald eagle, our national bird.

And, best of all, no license, fee, or special permission is required for looking.

But, as you will discover in the chapter to follow, there is, so to speak, more to looking at wildlife than meets the eye.

JACK DENTON SCOTT

·1·

SEEING

Seeing · 13

IT WAS a winter tree at dusk, stark as a surrealistic painting. An apple tree, with gnarled branches, bare twigs—just an old tree asleep in the cold. But something was going on there, as it always is in nature. The signs were obvious—if you knew how to look for them.

As my weekend guest came to stand beside me at the window, I played a favorite wildlife game with her. Pointing at the apple tree, I said, "What do you see?"

She looked intently. "A poor old tree without its clothes. No leaves."

"That all?"

She looked again, aware that I saw something that she should see. She stared for a long time, then shook her head. "Just a bare tree."

So I went through the routine: "What does a tree have in winter? No leaves. Twigs. Bare branches. Some of them gnarled, knotty. But rarely are the knots symmetrical. And they don't move."

My guest giggled. "Now I see them. Four birds. They just moved their heads."

I explained that these were ruffed grouse "budding," eating buds from the trees, as many as fifteen hundred in an hour. While many other forms of wildlife are on dangerously short winter rations, the ruffed grouse survive nicely with this technique. And at night they roosted in relative safety in that high tree.

We got the binoculars then, and as we noted the feathers around their throats touched with iridescence like a jeweled necklace, I pointed out that grouse also have built-in snowshoes. On both sides of their toes are rodlike projections which begin growing in October and triple in size by December. Projected along the full length of the toes, these comblike growths enable the birds to move more easily in deep snow.

The mention of snow reminded me of a sight last winter, when I discovered another survival technique of the resourceful ruffed grouse. I saw a depression in a deep bank of snow. Looking closely, I noted fanlike marks on either side and saw that the depression was actually some sort of entry hole in the soft snowfall. I poked it gently with my walking stick and almost had a heart attack. A grouse burst out like a flying missile, throwing snow into my face and aging me by at least ten years. I later learned that ruffed grouse often dive into snowbanks to keep warm. Here, also, they are wise about their safety: They don't dive in too deeply, so if any predator sniffs them out, they can fly from the snow suddenly, throwing a blinding snow spray into the face of the investigator as they bomb away on their stout, quick wings.

Thus four bumps on an apple tree, a hole in the snow, and the results from looking at them carefully. But you have to really "see." There is a whole world some of us do not observe and appreciate, because we do not watch wildly.

Watching wildly is a technique I was first taught by my friend Rao Naidu, a *shikari,* or professional hunter, in India's central provinces. I received an early lesson when we went to photograph the black buck, India's most beautiful and now one of her rarest antelopes. We drove to the area first to reconnoiter a *maiden,* a huge meadow bordered by jungle. We went in the morning, with the sun shimmering on the meadow that spread before us like a golden lake.

"The wild ones," Rao Naidu said in his soft voice, "look for danger when the normal is abnormal. They know what their territory looks like when it is undisturbed, and even if the intruder is a master of camouflage and stealth, by immediately spotting the deviation from the norm in terrain, the animal often saves its life. Primitive people in this area use the same technique when hunting, or when traveling where there may be danger from wild animals."

He made a sweeping gesture with his hand of the meadow before us. "I tell you now there are no antelopes here, but there could be and I'd wager that you would not see them. Study this place well. It is now normal. It is the way it looks when it is undisturbed."

There was a breeze. The grass in the meadow stirred; little waves moved across the area like ripples of water. The trees in the background stood as if cut from tin. There was a dead silence. I looked intently for fifteen minutes. Rao Naidu nodded, and we went back to camp.

We returned just before dusk the next day. There was still sun, but today there was no breeze, and the trees were casting shadows in long lines on the grass. At first glance I saw nothing. After I had looked carefully for five minutes, the meadow still seemed deserted. Then my intelligence joined my eyes in the search. In some places the grass was thicker; closer observation showed that this wasn't grass at all, but dun-colored rocks that I hadn't seen before. But I had looked well yesterday; there were no rocks. I passed this on to Rao Naidu.

He grinned. "Give you time and you'll learn to see. Those rocks weren't there yesterday. And they aren't rocks. They are black bucks, grazing with their heads down. They have heard us and they are remaining motionless, hoping we won't see them and will go away." He put two fingers into his mouth and gave a piercing whistle.

Immediately three-foot spirals poked up from the grass like a hidden band of men lying on their backs raising their swords—the horns of the antelope. The herd of eight black bucks—five males, three females—raised their heads and looked around, then suddenly took off in a bucking, angry-horse motion across the meadow and into the trees, their creamy undersides gleaming in the late sun, their tall, twisted horns swinging from side to side as they tossed their heads.

Separate the abnormal from the normal in terrain or objects, and learn to see the signs that are everywhere in nature—signs that tell you something is happening. This was the lesson that I had tried to pass on to my guest when I asked her to tell me what she saw in that apple tree.

In Kenya an Englishman talked about learning to see somewhat differently than Rao Naidu. They had the same viewpoint, but this hunter went into more detail. He was disgusted because I couldn't see a leopard standing in grass three hundred yards from us. The spots on the leopard broke up its body contour, hiding it in that grass as surely as if it had stepped into a hole and vanished.

I finally saw the cat, not because I had yet learned to watch wildly, but because we were staring at the right object at the right time, and a fly or some such insect apparently lighted on the cat's ear, causing it to twitch. My eyes moved down to the cat's—manic eyes touched with topaz. I often awake at night and remember them with a finger of chill sliding down my back. A leopard silently watching you from the grass is a sight not to be forgotten.

"We look but we do not see," the Englishman said as the leopard finally vanished in the grass, and we drove back to camp. "We're poor weak-eyed creatures, our eyes ruined by television and motion pictures, and by not properly using them."

This Englishman had given up civilization over twenty years ago in favor of the peace that his profession of hunter in Africa had given him. He was almost a native in outlook and could see a guinea hen in grass two hundred yards away.

"From my years spent here," he said, "I have learned that primitive man sees best. He can see extremely well without having his mind crammed with distracting thoughts. In fact, few civilized people know how to really see. We think we are seeing because we force our eyes to rest on an object. Less than ten percent of what our eyes rest on registers." He pulled the Land Rover to the side of the road and stopped to finish his conversation.

"The really serious watcher centers, targets his consciousness. By that I mean that the expert knows *he* must be *where* he is looking. All of him. Our eyes are not mere instruments recording, but they are also a means of outward projection of consciousness.

"To simplify, the eye and the mind working in harmony must separate the *ordinary* from the *strange*. To really see, you must empty the mind of everything—daydreams, envy of your neighbor, worries about the end of the world by nuclear explosion. You clear the mind, so you can project *yourself* out there where the eye is roving."

I never learned to see as well as he did, nor as well as Rao Naidu, but what I did learn has given me much pleasure. Quite often what my eyes rest on, I really see. And to double the pleasure of seeing, whatever you do observe in nature takes on added dimension if you research the subject and learn more about it, be it tree, animal or bird. Watching wildly is also a process of learning.

It is also a philosophy of life. If one sees, appreciates, and then takes the time and the trouble to learn more about the objects of these observations, this is the very essence of sound education, of adding valuable information to the mind-bank. Education, after all, is not only knowing, but knowing the questions and where to look for answers to these questions that pique and puzzle.

If you see something that fascinates you, find out more about it, be it peculiar cloud, unique sound, strange plant, or odd insect. By doing this, you are on the way to becoming a more complete person.

Fresh in my mind is a Maine fishing guide with whom I spent three days. He knew the names of only several common trees, none of the plants, and only a few of the birds in the wilderness where he had spent most of his life. Before I left, I wheedled him into going to a local library and reading some simple books on his own environment.

Libraries liberate the mind. Use them. So does nature. Observe it.

I also hasten to point out that the wildlife and terrain of India and Africa I mentioned previously are not necessary to practice seeing properly. They are mere examples from my own learning experience. Your own backyard is fine: Unless we are apartment dwellers, many of us during the winter months make a humanitarian sport of filling bird feeders and watching the action. But do we see enough? There is much more than bird-watching involved.

Among more than a dozen people talking about bird-watching at a recent gathering I attended, none had been lucky enough to see what I had under my feeder. I had noted three or four tiny holes in the snow near the feeder and watched carefully from my window, wondering what they were. It didn't take a great deal of patience, but you couldn't take your eyes from the scene, for it was a now-you-see-it-now-you-don't action, quick as an eye blink. Darting from a hole in the snow to the feed scattered on the ground for the sparrows and juncos, came a tiny, gray, mouselike shape—out of it, then back into the hole so fast that a bluejay feeding there struck inches behind it when it tried to spear it with its stiletto beak. The little creature was bold, too. And most of the birds on the ground didn't bother it. It came from the opening eight times while I watched. Much smaller than a mouse, it had what looked like half a tail.

When I checked my books to determine what it was, I thought it remarkable that it hadn't turned on the bluejay and eaten it. For that rodent taking advantage of my bird-feeding hospitality was a short-tailed shrew. Despite its size ($^4/_5$ of an ounce), it is probably the fiercest animal in North America. Possessing such a rapid metabolic system that it will starve in less than a day without food, the shrew preys on animals five times its size and eats more than twice its own weight daily. Paralyzing its victims with poison, it then devours them alive. I also have news for the biologists: If it can't get meat, the short-tailed shrew seems to do well with bird feed.

There is another cunning character that frequents the bird feeder. But, again, to catch it in the act takes practiced watching, and binoculars often help. In the fall most birds do not mate, so when they chase one another, it is because of competition or aggression. I knew that, so one day when I stood watching the feeder and an unusual bird came swooping in, frightening everything—bluejays, cardinals, finches, titmice, chickadees, and nuthatches—I wondered what it was. Smaller than a cardinal, with somber gray feathers and a black stripe down each side of its face, it sat on the feeder for an instant, then flew into a small tree perhaps a dozen feet away.

What I saw then makes even the violence on television pale by comparison. The first bird back to the feeder was a chickadee, a bold fellow who dove in, grabbed a sunflower seed, and started for a tree where it could crack the seed. As the chickadee left the feeder, the gray bird left the tree and came at it with bullet speed. The chickadee dodged and twisted and tried to outfly the attacker. An error. The gray bird beat the chickadee to earth with its wings and curved beak, fiercely tearing at its head. Finally picking up the battered body of the chickadee, the gray bird flew to a nearby barbed-wire fence, neatly impaled the chickadee's body on it, and proceeded to remove its head and eat it. Satisfied, it flew away. That was a shrike, commonly called a butcher bird. Not much larger than the songbirds upon which it preys, the shrike doesn't have the effective talons of most birds of prey. Its curved beak is its weapon, and it uses barbed wire or thorns, rather than its weak feet, to hold its prey while it selects choice portions, often just the brain, of its victim.

On the other hand, my neighbor, George Wheeler, saw an attack near his feeder that proved that a bird many think of as a villain can be a benefactor. George Wheeler is a country man who appreciates and respects wildlife and long ago learned to watch wildly, which by now we define as watching patiently and with precision, not just scanning. He has learned much. Recently when he was watching the songbirds at his feeder after lunch, a shadow fell, then vanished, then fell again. If George had been an ordinary observer, he would have had another look at the birds at his feeder, then gone about his business. But he isn't, so he patiently stood and watched. And he was surprised. He thought that a hawk had cast that shadow. But it was a large barred owl, disproving many an expert who claims that they hunt only at night. The owl flew from a nearby tree, gently as floating paper, and sat on the top of the feeder. A big brown fellow with vertical bars, it watched the birds beneath, then suddenly swooped close to the house, rising with a rat squirming in its talons.

Stored feed for the goats that are raised nearby apparently attracted the rat. The owl stayed two weeks, taking mice when it didn't see a rat, never touching a songbird or squirrel. Watching with a combination of eye and intelligence, George Wheeler found he had a friend that he hadn't even known existed.

All the foregoing might be called an essay on the eyes and how to use them to watch the world of nature in a rewarding way, to appreciate nature's perpetual theater and drama. The price of admission, however, is high. Patience, silence, and the ability to sit quietly for long periods are all required.

I hope that the reader will also be patient with my frequent intrusions into the pages that follow. But I pass on sights that I have seen and lessons that I have learned, looking through my window on the wild, to inspire you to open your eyes and look around you. It is an exciting world out there.

Look through your own window and you will find surprises and fascinating characters that you have seen often but never really appreciated. For example, in the chapter to follow you will meet one of the most common of all creatures—the earthworm. In proper perspective, both in the seeing and in the researching, the lowly worm is uncommonly interesting, with unusual traits, physical assets, and values few of us are aware it possesses.

· 2 ·

EARTHWORM

Earthworm · 27

IT WAS NOT an unusual sight. Many of us have looked out our windows on a late spring day and watched a robin in a light rain, standing as alert as a sentinel, listening for its underground prey, the earthworm, one of its main sources of food.

The bird's hunting technique is interesting. Sometimes the robin hops lightly, head cocked. Then it stops. It may continue with the hopping or switch to brief runs, then halt, always with the head cocked. Is it listening? Experts are divided in opinion, some believing that the robin actually sees the worm, others that it hears it underground. Whatever the radar, often after the series of hops or runs, the bird stops and makes a fast stiletto strike with its beak. And then, often, success. Slowly, skillfully, the robin pulls the worm from the earth.

Oddly, despite the patient and expert stalk, this day I did not think of the robin. I thought of the worm. We have all seen them in a garden and perhaps squirmed at their slimy, slippery appearance. Our hoe comes down sharp as a knife, cutting the soil, preparing it for the seedlings of flowers and vegetables.

There is the sudden white flash of underbelly, the frantic wiggle. The body separated by the quick blow slithers away, each half in a different direction. Both parts of that body burrow into the soft soil. The gardener probably didn't know it then, but that severed worm was more responsible for the later success of the garden than he or she was with all the hoeing and fertilizing.

Most of us know little about worms. We use them to attract fish to our hooks. What else do most of us know? I vaguely recalled that there are farms that sell worms by the thousands. To whom? For what purpose? For fishing, I assumed. But I wasn't sure. And would those halves of that worm that I had cut in two live? Regenerate? I didn't know that either.

Since that late spring day when I saw the robin pull the worm from the ground, I have learned that without that lowly creature we would have a much more difficult time raising food, thus even existing. This unpaid soil chemist, cultivator, plower, and irrigation specialist has also aided us in medical research.

Consider that this complex character lives successfully in tropical rain forests, mountains, gardens, rotting logs, or dunghills.

As we sleep, it comes to the earth's surface, fastens its tail in its burrow for safety and a fast retreat, and makes our world secure—by eating it. Consuming its own weight in leaves, grass, herbage, and earth every twenty-four hours, it prevents floods, stops erosion, and converts stone into soil.

Without the earthworm, it is said that it would take nature five hundred to a thousand years to produce one inch of topsoil. The worm can do it in ten years.

Originating millions of years ago, the earthworm has wiggled in a vast salvation army of 2,200 species across most of our world. Some appear in Australia and South America as seven-foot giants; one in Texas glows like a lantern; another, *Didymogaster sylavticus*, protects itself by throwing a four-foot biting alkaline spray from twenty cells; other kinds sparkle in shades of blue, gray, green, red, and purple. Measured by any yardstick, even the common four-inch earthworm most of us know is a fantastic creature with five hearts, two stomachs (one outside), and two sets of powerful muscles—one set that is circular, pushing the worm forward when the muscles are contracted, and one set that is lengthwise, propelling the worm's rear portion. Eight hooked bristles or feet, called *setae*, on each of the body's 120 segments spike the earth and glide the worm along.

Although it is eyeless and earless, sensory lenslike cells strung along its skin surface enable the earthworm to detect even slight degrees of light such as the deepening of dusk or faint dawn light. Vibrations as tiny as the hop of a robin or the feather-brush footfall of a field mouse instantly send the worm deeper into earth.

Lop off its head and it will grow a new one; cut that off and that one will be replaced. The tail? Slash it, and before long another will be wiggling.

Long fascinated with this phenomenon, scientists have discovered limitations: Cut the worm exactly in half and it does not become two whole worms. The half with the head will grow a new tail, but the half without the head dies.

But its limitations are few: Without lungs, the worm breathes through its skin, its damp surface taking oxygen from the air. Therefore it avoids sunlight, which slows this function. Heavy rainfall can sometimes be fatal; water in soil rapidly loses oxygen, and the worm can drown if it doesn't wiggle out soon enough. However, scientists have discovered that it can remain in waterlogged soil with water above for 50 days, live 137 days without food, and also withstand drastic dehydration, sustaining a 50 percent loss of body weight.

A built-in defense system fights this last enemy. During extremely dry weather it aestivates deep in the earth to sleep out a dry spell. It curls into a ball in a pocket of air in the ground. Before sinking into a torpor, it sprays mucus from special cells, painting a protective wall. By weatherproofing the surrounding soil it can remain this way for two months.

The worm has a choice of life-saving mucuses: One, exuded from back pores, emits an evil odor that discourages some enemies. Another makes it slippery to handle, forming a second protective coat over its already existing armor, a peculiar hornlike substance called chitin. This tough cuticle covers the entire body of the types known to most of us (the red family, *Lumbricus;* and the gray, *Allolobophora*), at times glowing with a startling iridescence. Sunlight broken up by reflection against hundreds of fine parallel ridges on the worm's chitin sometimes causes the worm's hard coating to flash like a jewel. But normally daylight isn't part of its natural world.

A creature of the night, of dark places, the worm usually leaves its burrow at dusk or on cloudy days when its skin radar indicates that the time is right. Whether it is surfacing to find moisture, to seek a mate, or for a change in diet, food is always there for the lucky worm. Unlike other animals that build burrows, it doesn't have to push the earth aside. It swallows it, nourishing itself with organic matter. As it eats its way through the soil, the worm also uses that marvelous mucus again to cement the sides of the tunnels, preventing them from caving in.

Sometimes submerging eight feet, it can gobble out an extensive burrow complex in four days, creating "pore spaces" that aerate the soil, permit the passage of gases and water, build drainage channels, and provide space for surface plants to sink and spread their roots. Experiments prove that without working worms, soil becomes soggy and structureless, has little resistance to water action, and erodes.

As a soil chemist, the worm has few equals. Calcium, nitrates, potash, phosphates, magnesium, and other earth minerals and chemicals pass through the test tube of its digestive tract and are broken down, enriching the soil with particle nutrients that small plants and seedlings can easily assimilate. At the same time, the earthworm is able to neutralize alkalies and acids. What it eats and digests emerges in little clumps of humus or "castings." Most average ground contains both the "red" earthworm that deposits castings *on* the surface, and the "gray" that casts *in* the soil; thus one builds topsoil, the other enriches the earth.

Its underground burrow work completed, the earthworm surfaces, fastens its flat tail inside its burrow opening, then sways to and fro, searching for food. Aided by the secure tail hold, at the slightest sound it snaps like a rubber band into the burrow. Omnivorous, it will nibble everything from dead insects to the tips of pine cones. If a leaf or a blade of grass is too tough, it coats it with a sort of saliva that eventually breaks up the food, or it buries it to decompose it.

German scientist Von Hensen, after studying earthworms for years, came to the conclusion that they have surprising intelligence. He watched them come to the surface and draw leaves into their burrows by rolling them singly or grabbing them by pointed ends to get them through the narrow space.

In 1926 Prague scientist Rudolf Malek watched a worm struggle to pull a leaf that Malek held tightly. The worm tried a new hold in several places, then finally gave it up as a bad job and found another leaf.

Tests conducted with two hundred red and gray earthworms showed they were highly selective. Given leaf mold mixed with grass, small plants, leaves, herbage, pine needles, cones, and twigs, the worms first ate the soft herbage and grass, which was easily digested, then the leaves, avoiding the tough veins. Pine needles and cones and twigs were quickly buried to age for future consumption.

The earthworm can be taught which way to turn. Dr. Rodabe P. Bharucha-Reid of the University of California tested it in a T-maze. One portion of the maze held sandpaper and an electrode that shocked, the other held moist earth. It didn't take the blind worm long to decide which way to go.

It isn't this small intelligence, however, but the large energy that makes the worm one of the most valuable of all animals on earth. Charles Darwin, who studied earthworms for forty years, came to the conclusion that they labored at incredible lengths to prepare the vegetable mold which comes to the surface in every moderately humid country. Darwin found that in much of England more than ten tons of dry earth yearly pass through earthworms and are deposited atop every acre. Sandy soil was also chemically converted into humus. Even rock is no obstacle for the earthworm. Exuding penetrating salts and acids, it can, in time, decompose it. Rock is solid soil. When it is broken and blended with vegetable ash, it becomes clay.

Many years after Darwin, aided by vastly improved modern techniques, English scientist W. J. McL. Guild came up with even more astounding facts in a two-year study of this infertile clay. In difficult clay pasture, earthworms churned up 11 tons of humus per acre a year; in excellent soil they managed to produce 24½ tons; and in rich tropical and subtropical areas, annual deposits of topsoil shot up to 100 tons an acre!

Pitting modern science against the earthworm, an experiment in Connecticut brought humbling conclusions. Good acreage was carefully prepared, the top five inches cultivated with a disc harrow to properly mix stubble and other organic materials with surface soil. This was then chemically tested with a comparable amount of earthworm castings. Result: The mighty worm produced five times as many nitrates, twice as much calcium, two and one half times more magnesium, seven times the amount of phosphorus, and eleven times more potassium.

In attempting to pinpoint scientifically how all this was possible, another soil specialist discovered that the actinomycete content of the soil (organisms between bacteria and fungi), which plays the major role in decomposing all organic matter into humus, multiplies *seven* times as it passes through the earthworm.

This amazing small creature works its earth-moving miracles by sheer numbers. A population of a half-million earthworms in an acre of normal ground is not unusual, and more than four million have been found. Each mature worm casts up about 8 pounds of humus a year. This mysterious underground activity can be pulled into clearer focus for each of us by comparison. Figure conservatively that 1 acre of good garden ground contains from 25,000 to 1 million earthworms, churning from 6 to 20 tons of topsoil a year. That acre of worms is worth 10 tons of expensive organic fertilizer a year and is superior to 2 skilled gardeners, each working twelve hours a day. Next time you unearth a worm, don't look with revulsion, behead it with your hoe, or capture it for fish bait.

Gently return it to the soil. It sets your dinner table, puts flowers in your garden, gives you the shade of trees, and helps supply lumber for your home.

With all this, the earthworm still finds time to take that dry-weather sleep and go away for the winter, heading below the frostline to gather with its own kind. Forming a huge ball, more than a hundred earthworms gather to prevent the evaporation of moisture from their skins. When the ground thaws, they separate and move upward. As the earth warms and their appetite is momentarily satisfied, the worms have another urge: the instinct to mate.

As with everything concerned with worms, mating is no ordinary affair. When they are just under a year old, in late spring or summer, food and caution forgotten, they head for the upper region, sometimes leaving the burrow to wiggle aimlessly atop the earth in their annual quest. They like someone their own length and, as they are hermaphrodites—all earthworms possess both male and female organs—they need help to reproduce.

Using their bristle feet to try out a prospective mate, the worms test for size. When they are satisfied, the two nestle, held together by the bristle feet and a sticky embrace of mucus. They cross-fertilize all night, looking like one large, fat worm.

A cocoon forms like a halter below the head of one. Eggs are deposited and fertilized by the union. The halter is finally slipped over the fertile worm's head, hardening, ends closing, and is carefully buried in moist earth. Four weeks later, from three to twenty young wiggle into life fully prepared to cope with its problems. They mature in three months. If they are not taken by birds, frogs, toads, shrews, moles, or humans, their life span can stretch as long as ten years.

But life span, it appears, is relative. In 1956 an earthworm colony, believed to have been buried for two thousand years, was discovered near St. Albans, England. Thought to have been there since the Romans entered ancient Britain, the colony was found by British scientists under a paved car lot, fifteen feet beneath the surface. While they were at home in a ditch that was filled in before A.D. 100, the remarkable worms were successively covered by beaten earth, gravel, tile, mortar—and many buildings—and the colony lived on.

The marvel of the worm seems endless. Dr. Edwin L. Cooper, assistant professor of anatomy at the School of Medicine of the University of California in Los Angeles, engaged in important research that involved the rejection mechanism, which is still an obstacle to heart and other human organ transplants. Cooper discovered that the mechanism in the earthworm is activated by special cells called phagocytes, similar to those in humans, which are scavengers that attack and destroy foreign grafts. The doctor has also found that certain cells and tissues change, age, and grow tumors in earthworms as they do in our bodies.

How many ways can the earthworm go on helping us? Who knows?

But I do know that now I never see a robin yank a worm from the earth without wincing, and I am careful when I hoe my garden. I also stop when I see worms struggling for life in a puddle of rainwater, lift them out, set them on dry ground, and am grateful for this incredibly beneficial creature that builds our world from the ground up.

3

MONARCH BUTTERFLY

Monarch Butterfly · 39

LIGHT AS PUFFS of thistle caught by the wind, they drifted past the window, gleaming in the September sun like bits of stained glass.

One lit on the window, wings fluttering. As I saw it close up and admired its beauty, I remembered another time when I was inspired to discover what the flight of these unique creatures was all about.

For the past several decades I have been fortunate enough to live in rural Connecticut, where the mad rush of the world mainly passes me by. Formerly I dwelt in the midst of almost two hundred acres. Today it is much less, but woodland still surrounds me, interspaced with several meadows, and the nearest neighbor is still almost a mile away.

When the pressures of civilization become a burden that I have trouble shaking off, I sit at the edge of a meadow and let the peace that is still there flow over me—a gray squirrel chattering in an oak tree, the clear caw of crows dropping from the skies, often the dainty step of a whitetail deer coming cautiously out of the woodland to a meadow.

This day as I sat, one of the most graceful sights I have ever seen in nature unfolded—a sight that was also the discovery for me of one of the most remarkable creatures that flies wild and free.

Without warning, I saw a tiny russet chipmunk leap up from the meadow grass at a butterfly fluttering above it. As if it was hurled by strong coil springs, the chipmunk continued leaping at the butterfly just out of its reach.

It was a ballet more stirring than anything I have seen performed on a stage: the little chipmunk in its twisting, determined leap, the butterfly rising ever so slightly with each leap.

That beautiful butterfly, vividly marked with orange and black, was a monarch, a fitting name for this royalty of its species. It was September, and that fragile creature was making its long migration south before winter arrived.

No, the chipmunk didn't get the butterfly. It always stayed just a little out of reach.

Fluttering unerringly on its incredible journey, that bit of ebony-and-gold gossamer, lighter than a bird's feather, would face many hazards and enemies before it reached its destination—if it did.

As I watched that chipmunk leap and leap again, and the monarch gain altitude and flutter on, I knew then that I would be a sentimental and devoted monarch-watcher for the rest of my life. I hastened home to find out all I could about this remarkable creature.

The story of the monarch is enough to humble even the most successful and determined of us. I never watch them make that fall flight south without a lump in my throat.

Naturalist Loren Eiseley even wrote an ode to the monarch. Entitled *Wind Child,* some of it went,

> Those red-gold drifters edged in black that blow like leaves . . . flickering evasively through the last sunlight

. . . blown to sea, lost to children, lost to enemies but beating on, beating on . . . speed fourteen miles an hour on a three-thousand-mile course to Mexico . . .

But too many of us look and do not see, and therefore do not feel. Nor do we take the time to find out what would make a flock of reddish-gold and black butterflies gather in flight, or that the flight is about evenly divided between males and females, and that a male determines direction and destination, using a unique chemical signal, a "pheromone," to assemble the butterflies.

The consensus of most viewers when they see monarchs massing in the sky: a great cluster of butterflies, mere insects that will soon die. That's not the way entomologist Jo Brewer sees them. "Without them and other insects," she observes, "plants would not be pollinized, soil would not be aerated or enriched, vines and trees would not bear fruit, and flowers wouldn't bloom."

Those beneficial migrating monarch butterflies do not attract our attention as do the haunting, nostalgic, wild cries of the geese. But their silent flight is, for me, more impressive than that of the big gray-and-black Canadas. The geese only have hunters and vagaries of weather to deter them.

It seems that the migrating monarch has just about everything against it, from leaping chipmunk to leaping small child. Its enemies are too numerous to list: wind, rain, frost, small animals, birds, other preying insects, and butterfly collectors, to mention just a few.

What about this amazing insect, the only butterfly among the 50,000 species, including the 750 that are native to the United States, that migrates?

How can a frail butterfly with a wingspread of only 4

inches, weighing only 1/50 of an ounce, fly from Canada to Mexico?

The "how" isn't definite, for much about the monarch still mystifies scientists, but the "why" is well known. They migrate because they cannot exist in cold climates, and in order to survive they must have nectar from flowers.

But the reason for migration is somewhat more involved. All butterflies belong to the order *Lepidoptera*, a huge clan of scaly, winged insects made up only of butterflies and moths. Moths, as most of us know, are not so brightly colored as butterflies; they fly at dusk or at night and are smaller than most butterflies. Also, moths have stringlike antennae, fasten their front and rear wings together during flight, and, while they are resting, hold their wings umbrellalike above them. The vividly colored butterflies soar about during the daylight hours, their antennae are thick, or clubbed, and at rest they hold their wings in a vertical position.

The monarch specifically belongs to the select *Danaidae* (milkweed butterfly) family that mainly inhabits tropical regions. Most members of this family lay their eggs exclusively on the leaves of milkweed plants. There are over a hundred species of these plants in only one genus, *Asclepias*, which provide a huge food supply for monarch larvae. But the plants cannot survive freezing temperatures, so the monarchs, *Danaus plexippus*, have the most important reason of all for migration: perpetuating the species. By using the milkweed plants in the north in summer, then flying south to find the plants during the winter, they are assured of a continuous food supply and, therefore, continued existence.

Why can they only use milkweed plants? Because by laying their eggs to hatch on such plants, the monarchs are ensuring built-in chemical protection from some enemies. Most species of milkweed plants contain varying amount of cardiac glycosides, a drug much like digitalis, which is used to stimulate weak hearts in humans. Ingestion of too much of the drug, however, causes severe vomiting.

How do the monarchs use this drug to their advantage?

It all begins with that most basic of all functions to perpetuate any species—mating. A male (males outnumber females slightly) scents a female from a distance and begins seeking her. Often, however, several males will scent the same female simultaneously and a contest begins with a half-hearted aerial combat designed to attract the female's attention.

Somehow the female makes her selection, and the mating —sometimes in the air, sometimes on a flower or the branch of a tree—is consummated.

Soon after this impregnation, the female begins laying her eggs upon the milkweed leaves. Only one egg, looking like a jade raindrop, is laid upon a single leaf, rarely more than four eggs to a milkweed plant. The eggs are sticky and easily adhere to the "nursery" plant.

During the summer a female can lay as many as four hundred eggs, which are the nucleus for many generations to come. For example, if only half the female's descendants live to maturity, she will be responsible for about eight million butterflies.

Three to five days after the egg is laid, a velvety black, yellow and white caterpillar chews its way out of the egg, begins eating the leaves of the plant, and literally never stops eating. The milkweed plant's toxic fluids enter the caterpillar without harming it, and are later transmitted to the adult butterfly that will emerge.

Birds that eat either caterpillar or butterfly become sick

almost immediately and begin violent retching. Once they have eaten a monarch caterpillar or butterfly, few birds ever forget the bitter lesson and never again touch a monarch in any form.

So effective is this protection from birds that another butterfly, the viceroy, mimics the monarch by hovering around milkweed, but does not lay eggs on it. The viceroy resembles the monarch but is smaller, and its colors are less vivid.

But there are enemies other than birds. One, perhaps the most deadly, is not affected by the milkweed poison—the large wasp. These insects hover over the milkweed, waiting for caterpillars to hatch. The wasps pierce the skin of the caterpillars and lay a single egg inside each body. The caterpillar continues to grow, but inside, the developing wasp slowly eats all but the most vital organs. The caterpillar completes its metamorphosis, but a healthy wasp, not a butterfly, emerges from the cocoon.

Scientists have estimated that during the first two weeks of life the normal monarch caterpillar, untouched by wasps, multiplies its weight by about five thousand and grows out of its skin five times.

The caterpillar is an interesting creature. Its body consists of three parts, each with a pair of short legs that end in claws. These correspond to the six legs of the butterfly that will be born. But the extremely mobile caterpillar has another ten appendages called prolegs, which it sheds before it changes into the pupal stage when it encases itself in a silky cocoon called the chrysalis.

Before the pupa stage, however, the caterpillar goes through a series of moults when it literally grows out of its skin. Before the final moult, when the caterpillar is about two inches long, it spins a sticky silken web on the underside of a leaf and fastens its legs to it. Twisting violently, the caterpillar releases its hold on the leaf with the front portion of its body and wiggles out of the old skin.

About fifteen days after that moult, the pupa begins to form on the larval skin which the caterpillar has carried with it since it hatched from the egg. The caterpillar moves from leaf to twig, seeming to seek a secure platform to which it can attach itself and begin its final spinning to form its cocoon. This must happen soon after this final moult because if it falls to the ground before attaching itself, the pupa may be destroyed, or the butterfly that emerges may be deformed.

The magic of metamorphosis now enters the stage when the caterpillar is about to change. After gripping its selected leaf or twig with its hind legs and then weaving a silken mat, the curving claws on the legs tangle in the thick center of that silk mat, called the button.

Now the pupa forms around the caterpillar's head, gradually growing to the tail, discarding the skin. Reaching the button, the caterpillar's claws retract and a sticky fluid fills this space. A barbed pupa stalk, also known as the cremaster, forces itself into the silk mat like a spear, and the new, now completely formed case—the chrysalis—where there once was a caterpillar, is firmly secured, enabling the pupa to whirl and dislodge the rest of the larval skin.

The resultant light green, gold-spotted chrysalis has no protective cocoon as the moth has and is actually a "resting," or a composing, case. Within it the startling and almost unbelievable metamorphosis from worm or caterpillar to butterfly is happening. In about twelve days the cells reorganize into a winged adult butterfly.

The warmer the weather, the faster the emergence of this "imago," and the sooner the chrysalis begins to split diagonally to release the wet, bedraggled, orange-and-black butterfly. In another miracle of nature, delicate beauty is created from the ugly caterpillar. The entire process has taken about a month.

Clutching a leaf or twig, the butterfly begins forcing its life-blood, a green fluid, into its wings, slowly inflating them. If it is a sunny day, that fluid hardens into a stiff, skeletal frame. When that occurs and the wings dry, the monarch launches itself falteringly into the air.

If the day is dark and damp, however, the monarch can hang suspended from twig, leaf, bark of tree, or rock for twenty-four hours before it is formed and equipped to fly.

And it flies amazingly well. Its swift, erratic flight confuses and discourages some enemies. But the real strength of that flight is that it instinctively takes the monarch where it can sustain the species for generations to come.

One banded monarch was clocked as traveling 43 miles in one day—about twice as fast as a person can walk. Another flew 1,870 miles, from Ontario, Canada, to San Luis Potosí in central Mexico. This averaged out to a consistent 15 miles every day. Some monarchs make an astounding flight from the northernmost edge of their eastern range, in Maine or Canada, to Mexico, a distance of about 2,500 miles.

For many years that flight posed a mystery.

It was known that after their migration south, monarchs gather in about thirty roosts in California, the most famous of them being in Pacific Grove on the Monterey Peninsula. Here millions tightly cluster in Monterey pines and eucalyptus groves, where both trees and butterflies are protected by law.

Such locations were well chosen by the butterflies. Just far enough from the sea so they would not be threatened by ocean storms and great winds which would destroy them, still the roosts were close enough to the sea to protect them from heavy frosts, which also can be fatal.

But where did the monarchs gather on the other migration route, in the east? They were seen in some places in Florida, but not in sufficient numbers to repopulate the eastern section of the North American continent. Where did they go? Flights were observed in Texas and along the Gulf Coast, but again there were no massive gatherings in these areas.

Scientists suspected that an unknown roosting place must exist in Mexico.

They were right. But it wasn't until 1974 that a Canadian lepidopterist, Kenneth Brugger, discovered probably the largest overwintering colony of monarchs in existence in the central Mexican mountains. One hundred million butterflies were gathered on forty-five acres of cypress, firs, and pines, their clusters so heavy that all the branches of the trees were bent and many broken. In some of the area the monarchs were packed against the trees in such numbers that only the outlines of branches could be detected.

U.S. biologist Lincoln Brower, whose team also later discovered the Mexican hideaway, asked himself the question: "How can it be possible for the butterflies to fuel their flight on the southward migration, survive several months with no sufficient access to nectar, then perform the northward migration?"

He and other scientists, such as L. A. Schroeder, believe they have finally found the answer to that mystery: The fuel comes from two sources: (1) During the larval stage the caterpillar can convert about 80 percent of the carbohydrate in milkweed leaves into fat that is passed on to the adult butterfly; and (2) the monarch makes many stops during migration to sip nectar from a great variety of flowers.

Perhaps the most knowledgeable monarch expert of them all, Fredrick A. Urquhart, professor of zoology at the University of Toronto, who has studied the monarch for over fifty years (and initiated an international identification-banding program), may have discovered one method by which the fascinating insect finds its way across uncharted, trackless miles.

He removed a single antenna from a number of migrating monarchs and studied the result. These butterflies were not deterred in making their normal migration. But when he cut off both antennae, the monarchs were confused, flying in all compass directions.

Biologist Brower offers a different opinion. "Individual butterflies that fly south in the fall," he points out, "are several generations removed from their ancestors that flew north the previous spring. Their annual return to the over-wintering grounds is thus a complex, inherited behavior pattern in which learning plays no role."

Those migration flights are sometimes so large that a single one may be fifteen miles long, darkening the sky like a huge storm cloud. In fact, one such flight passing over the Cascade Mountains in Washington prompted a forest ranger to report an approaching storm.

Another mystery was solved not long ago. Formerly it was believed that most monarchs reaching the northwest point of migration were not the same ones that began the flight in the spring, but members of a second generation. Monarchs with wing tags, however, proved that a single generation does complete that entire round-trip flight.

It still is a puzzle how second-generation monarchs make

their way along a migration route they have never flown. How do they find the same ancestral grove of trees to overwinter, sometimes even picking the same tree in which the generation before them roosted? And how do they select certain trees for swarming just prior to the migration flight, often the same tree each year? And no one has yet been able to explain why monarchs cluster.

Although some scientists believe that male monarchs gather the flight and keep it together on its way south and north by a chemical scent, that does not explain how single, free-flying butterflies make the continent-spanning journey on their own, as many do. Not all monarchs swarm before migration.

No monarch lives long enough to make the migratory trip twice. Nine months is the average life span. How *any* live that long is another mystery, flying on to their destination, wings tattered, spent, besieged by so many dangers.

Perhaps many of these questions will never be answered, and perhaps we should not question the gift of such beauty, or look beyond it, but just appreciate it and ponder its mysteries.

One naturalist said that in addition to its amazing ability to migrate those hundreds of miles against all obstacles, the fragilely beautiful monarch, the most attractive and abundant North American butterfly, also has the special magic for all of us of "making summer stand still."

·4·

BEAVER

SILENCE HAS sounds. In certain circumstances natural noises not only accentuate silence, but make one more aware of the solitude of the untraveled places—a brook running over rocks; a male cicada vibrating its wings to attract a female; a hawk whistling as it tests the wind with the spread tips of its wings; a bumblebee buzzing on a sleepy summer day; a rose-breasted grosbeak trilling high in an oak tree, aggressively marking out its territory with song. These sounds do not actually break silence but weave a meaning into its fabric, telling those of us who listen that all is normal.

There are, however, natural sounds that shatter silence. One of the most startling is an animal sound of alarm.

One summer mid-morning as I sat watching a kingfisher flash like blue flame across the water of a pond deep in the Adirondack wilderness, there was a sudden, crashing sound, not unlike someone breaking a barrel with a sledgehammer.

I must have moved as I sat on a knoll looking through an opening between trees, a "window" giving me a clear view of the pond yet partially camouflaging me. A beaver saw my motion and slapped the water with its powerful flat tail as a warning to other beavers. I saw it stop in the middle of the pond, lift its head, regard me steadily for a moment, then vanish beneath the water.

As I sat there, I dreamed like many another of the days long ago when Indians owned their own world and white people were beginning to realize that they could dominate that world and also make fortunes from the beaver's thick, lustrous fur.

The beaver does have this way of sparking the imagination and stirring up nostalgia. It and the buffalo are the animals that most symbolize our past.

Historians tell us that the beaver is partially responsible for the development of our country, pointing out that often the first contact the settlers made with the Indians was to bargain and trade for beaver skins. The high prices that trappers and traders received for beaver pelts induced them to explore deeper into the wilderness to discover new sources. Trappers and explorers blazed trails that were then traveled by settlers, the new continent's horizons began to narrow, and the settlements grew. Fur centers became the cities of Albany, New York, and St. Louis, Missouri. Trading posts were converted into army forts.

But as I learned more about this unique, semiaquatic creature that spends half its life in fresh water, the strongest feeling I have every time I see one is not a sense of history, but admiration. The phrases that have enriched our language—busy as a beaver, eager beaver, among others—are inadequate when they are used to describe the remarkable actions and accomplishments of one of North America's most fascinating wild animals.

I am acquainted, however, with two people who weren't fascinated. One bought an old saltbox house in Connecticut and settled in with his family to enjoy the salubrious country life. Fate and a pair of beavers had other ideas. A small stream meandered through the property several hundred yards behind the house. Shortly after the family moved in, two beavers took possession of the stream, their obvious purpose being to build a pond. My friend didn't want a pond. But the pond grew quickly. It was stagnant and attracted hordes of mosquitoes. As fast as the beavers built the dam to create the pond, my friend just as quickly tore it down. The beavers immediately rebuilt it. This went on for two months. Weary, welted with mosquito bites, my friend gave up in defeat. He now had a pond whether he wanted one or not. His wife suggested he turn to the state for help.

This he did, and the state sent a trapper, who finally managed to live-trap the beavers and move them to another area.

The other friend, a retired admiral, had a creek bubbling through his property. A man of good sense, the admiral wanted to keep his property natural, not convert everything into lawns and pebble drives. He wanted trees. So did a large beaver. For its pond-building, the beaver went through copse after copse of young trees like a crew of lumberjacks working to create a tree-felling record. Dismayed at what was happening, but a staunch believer in individual accomplishment, the admiral finally made a tough decision: He would trap the beaver. After three weeks of frustration and planning, the admiral, an expert at outmaneuvering on and in the water, got the beaver—reluctantly. He admired that industrious beaver, an animal of heroic size weighing about sixty-five pounds.

Actually, beavers never seem to stop growing. If they are well fed and very old, they can weigh over a hundred pounds (the record is 110 pounds). In captivity they can survive almost twenty years, but in the natural world ten is normal. In the wild beavers usually don't live long enough to attain enormous proportions. The limit in length is about four feet and the average weight is forty pounds. Fifty is large and seventy possible. But beavers had an ancestor, the prehistoric *castoroide*, that realized everything the modern animal merely suggests. Shortly after glaciers ground out the world we know, perhaps a million years ago, that beaver—the size of a bear—lived in North America.

The present beaver (the world's largest rodent, except for the South American capybara) is large enough to cause consternation when it starts its dam-building operations in places where dams aren't wanted by people. In Hampden, Maine, a colony of beavers built a dam near the mouth of a culvert that carried a stream under Canaan Road, flooding fifty feet of roadway and hundreds of feet of land on either side of the road. It took six months and the best brains in town to finally outwit the beavers. Two beavers in Michigan dammed a culvert underneath some railroad tracks and sank a South Shore and Atlantic Coast diesel train. Another pair flooded a church in Maryland. Not long ago the Wisconsin Conservation Department received over five hundred complaints about beavers that had built dams in the wrong places. The state spent $50,000 dismantling fifteen hundred dams.

Beavers bulldoze right through the confining influences of civilization. One man, to defeat the beavers that were building a pond that threatened to destroy his cornfields, dug holes in their dam and set six steel traps. The beavers used the traps to help plug the holes.

Why do beavers create ponds? In order to live, to protect themselves, to store food, and to construct lodges. Beavers must have water at least two feet deep available throughout the year. They use this water as an escape hatch, a road to transport food and home-building materials, and a storehouse. But to have the water, they must make the dam.

Usually alone, but with its family if it has one, the beaver collects materials—stones, mud, driftwood, sticks or logs from which it already has eaten the bark (its favorite food). It even takes advantage of litterbugs, securing into its dam discarded beer cans and bottles. I have seen a beaver dam in New Brunswick, Canada, replete with a magnificent set of moose antlers, artistically arranged into the dam wall.

Everything is done methodically: Limbs and trees of one length are compactly laid together, on the downstream side, parallel to the stream's course. The log wall is weighted and cemented into position with mud, stones, plants, and shrubbery. At first when the dam is completed, water seeps through, but the master builder efficiently stops this by plastering mud on the upstream face of the dam. To waterproof it further, it guides debris and great waterlogged clots of leaves against the upstream side.

The dam can be of simple, straight construction or twisted and turned, and can extend 5 feet in height and 300 feet in length. But those are ordinary dams. One was found on the Jefferson River in Montana that stretched over 2,140 feet. New Hampshire holds the record, with a 4,000-foot dam that created a pond that became a small lake holding forty beaver lodges. Dams are so well built that after they have set for a while, an adult, a horse, or even a Jeep can pass over them without harming the walls.

Beavers, however, don't build dams just to keep themselves busy. If there is enough water without work, they take advantage of it. Some live on ponds or rivers where even at low level there is plenty of water for their needs. On a stream too large or swift to construct a dam, beavers build a burrow in a bank, locating it above the high-water mark. But even these "bank beavers" are good builders that can construct a room three feet across, two feet from floor to ceiling, and connected to the stream by a tunnel that is sometimes as long as fifty feet. Astutely designing their lodges so the lower end opens under the level of low water, beavers can move to and from this home without exposure on bank or water.

But the majority of beavers live in "lodges," which usually demands that a dam be built. The lodge may be near or away from the dam, on shore, on an island, or in shallow water. If the beaver decides on shallow water, it builds an island protruding above the pond, keeping in mind that it needs access to deep water so it can have a channel open under winter ice. To start the lodge, it assembles limbs and sticks and creates an oval with at least one side toward the water. The beaver constructs the walls of logs, mud, and branches to a height of three feet, built inward, then roofed. As long as it lives in the lodge, the beaver keeps it in better repair than most people do their homes, adding limbs and peeled poles and cementing with mud when necessary. The resultant structure can be seven feet high and nearly forty feet in diameter.

In the wild I have quietly watched this remarkable animal doing a repair job. A huge armful of mud held against its chest, balanced by its tail, walking like someone about to wallpaper a room, the beaver climbed up the side of its lodge and neatly plugged a hole near the top. It came back down like a child on a playground slide and *ker-plopped* into the water, mixing work with pleasure. I liked the beaver then and like it even more now. It is probably the most continuously interesting animal I know. Surely it is the most industrious.

The lodges are comfortable one-room abodes, five feet

wide, three feet high, large enough for a person to get into.

There are tales of Indian fighters escaping from red warriors by hiding in beaver lodges.

The beaver was probably the first real engineer. While humans lived in holes in the ground or climbed trees, beavers were secure in their island lodge. Surrounded by water, if they were disturbed they could dive and swim to safety. Historians believe that the protective moat surrounding the castle was an idea people copied from beavers.

The beaver's life (dam, lodge, food) depends upon its tree-felling ability. It can topple a five-inch-diameter tree in three minutes and cut down a hundred-foot-tall tree with a five-foot circumference in three nights. It prefers to work alone, but for an especially large tree, often a family working in shifts will fell it.

After carefully selecting a tree, the beaver sits up, bracing itself with its wedge-shaped tail, and cuts a notch head high on the tree, then moves down three inches or so and cuts another. Now it rips out the wood between the notches, using just the lower incisors, the upper teeth giving the necessary purchase. The beaver cuts small trees from one side, medium trees from both sides; large trees it girdles. Rarely stopping to rest, chips flying, the beaver keeps at it until the tree starts to creak and sway slightly. Then it slows the work and proceeds carefully. As the tree begins to topple, the beaver races for the pond, creek, or stream, dives and exposes just its head in order to listen and watch for enemies (otters or large carnivores) that may have been alerted by the noise of the falling tree. If danger is sighted, the beaver slaps the water with its tail, and all beavers within earshot scoot for safety.

When things are secure, the beaver waddles back to the

felled tree and cuts it up into two- to six-foot pieces. It drags a length over the ground to the water and floats it across the pond. In the autumn the beaver wears deep paths to the water hauling logs when it is hard at work gathering food and repairing dam and lodge for the winter ahead. It always uses sloping land to advantage to slide logs to their destination.

If the pond is some distance away, the beaver digs canals to float the logs. These canals are channels that are two or three feet wide and a couple of feet deep, and often just under a thousand feet long. This amazing animal sometimes plans these labor-saving canals on three levels—the lowest is filled with water backed up from the pond; the upper levels are designed to trap the flow from springs.

The beaver selects dam sites carefully, when possible with aspen (which we don't use but which is the beaver's favorite food tree) nearby and along streams that do not dry up in August or wash out in March. The dam also usually backs up onto a pond that is deep enough to prevent the beavers from being ice-bound in their lodge in February.

Beavers do not hibernate; thus, they must store a plentiful supply of food to last through the months when their ponds may be thickly iced. One method of storing, besides stocking the lodge, would seem to require foresight. Logs, often aspen, are buried in mud in water below the freezing line, or they are deeply sunk, weighted with heavy stones. (Beavers are extremely powerful and have been seen shouldering aside 40-pound rocks.) When additional food is needed, the logs are towed underwater into the lodge, where the bark is shredded and eaten.

It is commonly believed that a beaver has the ability to fell a tree in exactly the position that it wants it. But nature helps. Land along water often slopes. A beaver working on a tree on this land makes its cut higher on the trunk on the uphill side than on the downhill, thereby felling the tree toward the water and avoiding extra hauling overland to the lodge. Also the side of a tree that is toward the water gets more sun and therefore grows more branches. This weight also pushes the falling tree toward the water.

In addition to this "tree-sense," nature has endowed the beaver with impressive physical assets. It has four 3-inch-long, orange incisors, two above, two below, which are forever growing and self-sharpening. In addition to these tree cutters, it has sixteen molars to grind its food. Its 6-inch-long, fully webbed hind feet enable it to swim five miles an hour and travel underwater nearly as swiftly as a fish. The unwebbed forepaws, almost as controllable as the human hand, are closed into fists and held against the chest while swimming. Flat, scaly, black, about 15 inches long and 6 wide, the tail is a rudder, an oar, a diving plane, an instrument of warning, a mason's trowel, a fifth leg. When the beaver stands to cut down a tree or plaster the lodge, its tail balances it upright. And when it submerges, special flaps in its nose and ears close tightly, automatically opening when it surfaces. The eyes are set so that transparent lids (through which it can see underwater) close and seal them tightly. Its lips are so flexible that they can fit snugly behind the chisel teeth. With this arrangement a beaver can cut and even chew underwater without getting a mouthful. (A special gland in the stomach exudes an enzyme which converts woody material into food.) Especially large lungs and liver enable the beaver to store air and oxygenated blood long enough to remain submerged in water for fifteen minutes and to permit it to swim a half-mile underwater without coming up for air.

One of the assets that the beaver can be proudest of is its

coat, which varies from blond to black and is so thick (with a heavy, woolly undercoat) that even in water skimmed with ice, the beaver is not uncomfortable. I have seen a beaver come out of winter water and sit on a bank and calmly comb itself. Using the split nails on the two inner toes of its hind feet, it carefully distributed the oil from its anal ducts into its coat, seeming to deliberately waterproof itself for the next swim. (Ducts also lead from two 4-inch scent glands or "castors," which store a rich yellow oil, castoreum, that the beaver uses for marking and claiming its territory by mixing this scent-claim into little piles of mud.) The beaver even uses those split nails with which it distributes the oil to clean its teeth.

It was that well-cared-for coat that almost destroyed the beaver. In the early days of our country, its pelt was more avidly sought than gold. Beaver skins, which founded the vast Astor fortune as well as many others, were used as barter (a dozen skins purchased a rifle for a trapper or an Indian; one pelt was worth a pound of tobacco, eighty were worth $500 in a time when daily wages were 50 cents). The Hudson's Bay Company in Canada traded blankets and rifles for pelts and then sold three million skins to England in little more than two decades.

With such an assault, by 1928 the beaver was nearly extinct in nearly every state. From an estimated total of sixty million when the pilgrims first landed, only a few colonies were left in inaccessible areas. It was the story of the buffalo and the passenger pigeon all over again. Our greed had nearly wiped out one of our important natural resources.

Finally some states were prodded into action by conservationists—but only after it was pointed out that with the absence of the beaver, wells were going dry because water tables had dropped. And without the restraint of beaver dams, hundreds of thousands of tons of topsoil were washed out to sea.

By then very few beavers were left. Most that survived were in the isolated Northwest of the United States and the far north of Canada. These were live-trapped and used for restocking. The comeback was slow but sure. For example, in Alabama in twenty years the beaver population grew from 500 to 200,000.

Today less than fifty thousand skins are sold yearly in all of the United States. The beaver exists and is protected in every state; some permit trapping, but with controls, licensing, and limits.

Modern conservationists have also discovered that beavers and their dam, despite some road flooding and other inconveniences, are our friends. Old beaver dams slowly break up when they are abandoned. They drain and leave extremely rich loam sometimes as much as a dozen feet deep. Experts believe that this process, repeated constantly since before the beginning of civilization, has created much of our fertile farmland.

Agronomists believe that the main reason that Vermont has such poor agricultural regions is the hundred-year absence of beavers there. It has been proved that by parachuting beavers into areas where streams flood, the resultant dams, acting as a natural form of flood control, slow water runoff, conserve spring and early-summer rains and reduce soil erosion.

Beaver dams also create excellent habitats for waterfowl and wading birds, otters (ironically, the beaver's worst enemy, especially of its young), mink, and raccoons.

Geologists have estimated that it would be the rare stream north of the Mexican border that has not been "beaverized" by having its surrounding land enriched by these superb, beneficial engineers.

Freed from the danger of trap and gun and protected as a national asset, the beaver is once again firmly established on the wildlife scene. Once there were thirteen subspecies of the beaver in North America, their differences marked only by range, size, and color. Today, because replacement stock had to be obtained wherever it could be found, subspecies were introduced into areas where they had never been before. These have crossbred with the beavers that were there, resulting in difficulty in accurately identifying any subspecies. As a result, now *Castor canadensis* is just a beaver, not an Acadian, Carolina, Labrador, or Woods beaver.

And being beavers, they are well-organized, versatile colony creatures that don't exist on the bark of trees alone, as lumber lobbyists and uninformed amateur naturalists would have us believe, but eat most marine growth, grass, roots, all kinds of sprouting greenery and shrubs, and even have a fondness for pond algae. Unfortunately, the beavers that have successfully moved back into suburban areas also like flowering dogwood. But many residents agree that they'd rather have fascinating beavers than beautiful blossoms. And with some care, they can have both.

Firm family animals that like to keep their offspring with them longer than most wild creatures, beavers may be most interesting to watch, but they are all business, rarely frolicking or playing. When they aren't working on a dam, a lodge, or their food supply, they are educating their family, swimming beside the "kits," diving with them, racing underwater, leading them to food sources, pushing them in and out of the lodge, alerting them to danger.

The family begins when a pair of beavers matures at two years and mates, probably in the lodge during the months of January or February. The young are born four months later, in April or May, on shredded wood used as bedding in the lodge. Before she gives birth, the female evicts the male from the lodge. If there are yearling kits around from the year before, the male takes them with him and finds an abandoned lodge to stay in until the newborn young are weaned at two months of age.

Beavers have only one litter a year. They run from two to six (four is the average), all births taking place in about sixty-eight hours, with the female sitting upright, her flat tail forward. The young, born with their eyes open and fully furred, are about fifteen inches long, including the tail, and weigh just under a pound. By the end of the first day, the young can swim with the mother, even underwater.

A colony of beavers usually consists of a male and a female, the two-year-old young which remain until a new pregnancy begins, and last year's kits. Rarely does a family number more than twelve. The two-year-olds finally are driven from the lodge to make way for the new kits and to give last year's young space and food to grow.

If the two-year-olds don't leave the lodge voluntarily, they are not only driven out of the lodge, but away from the area itself. The parent beavers are not only protecting their food supply by preventing overpopulation; they are also avoiding inbreeding, which would weaken the colony and, eventually, the species.

The two-year-olds now go off to find mates, build lodges, and establish colonies. Naturalists have watched these new mates try clumsily and unsuccessfully to build dams and lodges. They have seen them stop work, leave, then return with mature beavers that repaired their amateur efforts while the younger animals helped and learned.

With such teamwork and cooperation, instinctive or not, no wonder the beaver is considered among the most successful of all wild animals.

·5·

BADGER

WE, AS A SPECIES, tend to have a problem observing and assessing the natural world. We generally invest wild creatures with human characteristics, create good guys and bad guys, actually see them in our own image. A complicated word, "anthropomorphizing," describes this mistake of giving wild animals human values and making them fit our standards of behavior.

For example, most of us consider the African lion noble, the traditional symbol of kings, our modern trademark of majesty. We believe that the lion is a skillful and humane hunter, a brave protector of pride and progeny. We also are convinced that the night-skulking hyena is a cowardly creature that feasts on the dead and lives off the leavings of other animals.

Many of us think of the wolf as vicious and occasionally human-eating; the gorilla as a fearsome King Kong; the crocodile as a cold, sinister reptile, a sort of animated log, all appetite, without brain or feelings.

These beliefs have little basis in fact. Consider: Lions prefer to steal food from other predators rather than hunt it.

In fact, they get more than half their food by scavenging prey killed by other animals or eating animals that have died of disease. And, if food is difficult to obtain, adult lions will eat all they can find and leave their cubs to starve to death. If they do hunt live prey, they often bungle it, rushing too soon and misjudging distance.

Just about everything we believe about the hyena is incorrect. Its main prey is not dead animals but healthy, adult wildebeests and zebras. It can even skillfully kill formidable half-ton African buffalos. The hyena, which also kills more quickly and humanely than the lion, does eat carrion, but not as often as the "king of beasts."

Wolves live in highly developed social groups, rarely fight within the pack or kill unnecessarily, and are only fair hunters, mainly selecting the weak and diseased, thus benefiting the prey they hunt.

Wild gorillas are shy, gentle vegetarians that would rather munch on a stalk of wild celery than attack a human; and crocodiles hunt cooperatively, share their food, and live in social groups. They actually are graceful, gentle mates and are protective of their young from the moment of birth—the female parent scoops the newly hatched offspring from the sand and carefully places them in her fanged mouth, carrying them to water and to life.

Some animals, however, do make it difficult not to use human terms to describe their characters and their actions. For example, one animal, burrowing beneath a heavy wooden platform, raised it inches above the ground. This may not sound like much of an accomplishment, except that platform held a horse and its rider. The animal was a badger, for its size one of the most powerful of creatures.

Again, illustrating its bulldozerlike power: a badger, its

right foot caught in a steel trap, dug a cavity four feet deep and four feet in diameter in an attempt to free itself. Another pair nearly closed a Midwest airport by undermining and buckling a major concrete runway.

With documented feats such as these, could anyone be criticized for calling the badger the Samson of the animal world, as one writer did?

No. But it is more effective to describe an animal by its own actions, rather than picturing it in human terms.

Also, it isn't fair either to us or to animals to use untrue statements and exaggerated claims, as is the case with the lion and some others.

Not much is exaggerated where the badger is concerned. I had heard many stories about this unusual animal's digging ability, some of them so extreme that I remained skeptical for many years—but only until I went on a marlin fishing trip to the wild region of Baja California.

Off the boat at dusk, I was stretching my legs in an open area of land when I surprised a badger. Shaggy, squat, heavy-bodied, and short-legged, it stood about nine inches high and was yellowish-gray with a white stripe extending from its nose over its head and fading back to its shoulders. It had vivid white cheeks, a dark face, and a black spot in front of each short, broadish, upright ear. Its feet were black, its underside and six-inch tail a drab yellow. It looked about three feet long. It cannot be confused with any other mammal in North America, and is one of our most remarkable wild animals, both in appearance and in performance.

Shy and quick, it gave me just about enough time to have this fast look to identify it. I had seen badgers before in the West. This one, caught in the open, without any cover, suddenly vanished before my eyes. I told myself that it must have found a handy burrow. Walking to where it had been standing, I saw a pile of freshly excavated earth. In an incredibly short time that badger had dug itself underground so deeply that there wasn't even a tremor in the soil where it obviously was still digging downward. The ground was somewhat damp and softened after a recent rain; still it seemed an amazing feat of strength and engineering. I have since learned that this isn't amazing at all, but standard operating procedure for the animal when it is trapped in an open area.

I wished then that I had a shovel so I could dig down and see exactly how deep that badger had dug in such a short time. Fortunately, I didn't. Naturalist William Curtis, studying coyotes in California, wrote that he had the same experience and tried to dig the badger out. He and a friend dug straight down for fifteen feet before giving up in exhaustion.

Although they are mysterious, wild creatures few of us know, badgers range all over northern and north-central North America; from the Great Lakes west to Nebraska and northeastern Mexico, Baja California, the Pacific coast, over more than half of the western United States, southern Idaho, British Columbia, central Alberta, and Saskatchewan. In recent years it has been reported that badgers are extending their range eastward. They have been seen in New York State and New England. It is thought that these animals were probably released or escaped from captivity; or they were lured east by logging, which encourages and expands the rodent population, a favorite badger food.

Badgers are nocturnal and seem shy to us because only occasionally are they out in the morning hours. But they are courageous creatures. One eighteen-pound mother with young did battle with a two-ton automobile, believing it an enemy. She was defending her offspring from the car that had stopped three feet from them.

This courage is the hallmark of the entire clan. Naturalist Victor H. Cahalane claims that because of the badger's fierceness few animals will attack it; that a dog pack has been known to best one, but that a badger can defeat and even kill a dog four times its size. Often, though, the badger will attempt to bluff its way out of danger by hissing or snorting loudly and bristling all the hair on its back, making it look about twice its normal size.

Its fighting accomplishments are possible because of three assets: courage, strength, and physique. The badger's skin, which is so loose that it appears to belong to another animal, seems several sizes too large, making it almost impossible for an assailant to get a killing grip. The badger simply twists around in that flexible armor and bites back. Its jaws are so strong and its teeth so sharp that they can easily penetrate even the toughest hide.

Two-inch claws, sharp and tough from constant digging, are as lethal at in-fighting as honed knives. It takes good care of its claws. A badger has been seen sitting in the morning sun meticulously giving itself a manicure, using the claws of one foot to clean those of another. Protected by exceptionally thick neck muscles and shoulders, bowlegged and pigeon-toed, the chunky badger looks like the fighting machine that it is.

At about the turn of the century, a "sport" was built around that quality of the badger. The animal was captured and placed in a barrel. People gathered and bets were placed among those who thought they owned able dogs. The idea was that the dog would pull the badger out of the barrel and not be badly mauled itself in the process. Few dog owners won any money.

William Curtis wrote about a famous fight between a champion English pit bull terrier that had defeated all comers in its section of California and a randomly selected badger. The match was widely advertised, with hundreds of spectators paying admission to see the "fun."

The fight soon became a legend. The big pit bull lunged against the small opponent and fiercely took its famous hold. But there wasn't enough neck to grab securely, and the badger's head was too fast and too full of teeth. The dog then tried the body. Sliding around in its loose hide, the badger ripped with its claws and tore its opponent several times with its teeth before the pit bull decided that this was the wrong tactic.

Thus it went for two hours, until both were bloody but unbowed, and no judge or spectator was willing to declare a victor. They finally let the brave badger go, and in its usual fashion it vanished underground. The pit bull could barely move. It died in a few hours.

With that staying power against unfair odds, and built like a clenched fist, the badger does not give the impression that it has speed or agility. It is supposed to be a slow runner and a poor swimmer. Naturalist A. Brazier Howell, however, surprised a badger at a beaver pond and was astounded to see the animal whirl, then bound away in ten-foot leaps. Another naturalist saw a badger gracefully dive into and swim across a small lake almost as agilely as a muskrat. Still another saw the squat animal run down a speedy squirrel.

That must have been a hungry badger. Normally it is not necessary for it to depend upon speed to obtain food. Truly omnivorous, it will eat fruit, nuts, and grasses, even insects, worms, and mollusks, and devour a nest of newborn squirming field mice.

This catholic appetite has puzzled some biologists, making them wonder if the badger is somehow immune to rattlesnake venom. In the Birds of Prey Natural Area, a 33-mile,

31,000-acre strip of land flanking both sides of the Snake River Canyon in Idaho, which holds one of the largest badger populations in the United States—about ten per square mile—a favorite food is rattlesnakes. The badgers have been seen killing and eating the snakes, apparently without any ill effects.

They also are among the few mammals that can successfully dine on porcupine, an animal whose coat of quills discourages most would-be enemies.

Some naturalists believe the badger's skill in catching its prey has created a curious companionship. The Mexican name (of Indian origin) *talcoyote*, "like a coyote," suggests a relationship between the two animals. Coyotes have been observed loitering around a badger busily excavating, until wood rats were flushed from their hole. Then the coyotes dashed in and caught the rats. This watching-and-waiting technique of the coyote has been documented many times. But one badger was seen switching roles, following a coyote as it chased a cottontail rabbit into a hole. The coyote couldn't dig the rabbit out and did not have the patience to wait until it emerged. The badger stayed until the coyote left. Then it promptly dug its way into the hole and secured the rabbit.

The entire underworld is the badger's prey ground, providing everything from a morsel of mole or a tender young rabbit to a spicy supper of sleeping skunk. Badgers aren't gluttons, though. They'll eat the smaller items and bury the larger portions for later meals. So they dig to eat, dig to save for another day, and dig to build a home den.

English naturalist Ernest Neal, one of the few to make an in-depth study of the badger, wrote in his book *The Badger* that the animals' home tunnels penetrate surprising distances. One was sixty yards from the nearest exit. Hunters sent a small terrier into another badger hole, or set, as the multiroomed tunnels are called. The technique: the dog darts at the badger, in theory keeping it from digging deeper or from escaping until the dog's master can capture it. The dog didn't return in the normal time. Later they saw it surfacing on the other side of the hill. One hundred yards away! And no badger in sight. They thought that one set of tunnels had probably led into another. But they weren't certain. No one is ever certain about a badger.

One thing, however, is certain. If a badger is chased, once it gets into a burrow it is safe. The technique for securing this safety is as sound as slamming a steel door behind it. The badger extends the tunnel in any burrow it pops into fantastically fast, packing the earth so solidly behind it that an observer even saw a powerful black bear fail to dig one out.

Naturalist Vernon Baily describes this remarkable animal as having the stubborn, never-say-die characteristics of bulldog and dachshund, the fighting ability of the weasel, the strength of the bear, the digging talent of the mole.

Ernest Thompson Seton, who sketched it many times, thought the badger looked like a small bear that had been flattened, colored silver gray or drab brown, then adorned with black and white marks on the head. Naturalist Ernest Neal has described some badgers as looking like a cross between a teddy bear and a panda.

Exactly who and what is this unique animal? Probably the major reason most animals don't stand a chance of outbiting the badger is their clan membership of *Mustelidae*. One outstanding asset of that family is the manner in which the lower jaw is locked into a long cavity of the cranium, giving the animal the ability to maintain its grip with great tenacity. In effect, once the badger clamps down, its teeth are virtually frozen into place. In addition to other attributes, the badger has a keen sense of smell, acute hearing, and marvelous reflexes. Its eyesight is not especially good, but it catches movement of any kind quickly.

Taxidea taxus, the American badger, belongs to the complex weasel family, which includes not only the large tribe of weasels itself, but martens, minks, otters, skunks, and the formidable wolverine, a close cousin of the badger.

It is a sophisticated group occurring in many parts of the world in six different classes or genera. *Meles*, the British badger that was discovered to have lived as far back as the Middle Pleistocene era, 250,000 years ago, is directly related to four other distinct species that appear in Japan, China, and Tibet. *Arctonyx*, the hog badger, is found all over Asia; *Mydaus*, including the stinking badger, inhabits the mountains of Java, Sumatra, and the Philippines. *Helictus* is made up of the five species of ferret badgers found in Formosa, Borneo, Burma, and the Himalayas. Another, *Melogale*, also is Asiatic.

Our North American badger, the sixth genera, has six subspecies, which differ only in size and color, according to their range. For example, the subspecies in our northern plains grew a long, silky platinum coat in the spring. For many years trappers received an excellent price for that hide. But wearing it was a fad, since the fur was not durable and was finally no longer sought after or valuable.

The so-called common brown badger, the most prevalent, also was avidly hunted. Trappers received eighty-five dollars a pound for its stiff, durable, yet soft bristles, which made the best shaving brushes. Progress, in the form of synthetic bristles, brushless shaving cream, and electric shavers, saved this species from extinction.

Cowboys and cattle owners also shot badgers whenever

they could, claiming that their burrows were dangerous because horses and cattle stepped in them and broke legs. It was an exaggerated claim, and the badgers proved to be better at vanishing than the people were at shooting, so they survived.

Also, conservationists came to the rescue, proving by experiment and research that badgers are so skillful at digging out mice and rats that they are an important factor in checking the destructive rodent horde which causes $700 million in damages in the United States yearly.

This unusual animal also has added a colorful word to our language: Probably taken from its fighting tactics, to "badger," worry, or harass is a metaphorical derivative in active use since 1794.

The badger's life-style enables it to live almost anywhere where there are burrowing rodents and soft earth. The vital stage of reproduction begins for each generation when the male, at the age of seventeen months, apparently goes beserk.

Once its scent has snared her, the male begins prancing around and purring, groaning, and yelping. Then it performs a series of leapfrog jumps over the back of the female. Next it somersaults, then rushes in and wrestles with her. These are clownlike actions, but with an ulterior purpose: to excite the female, as a prelude to mating. If the female has been impressed and is willing, the male and the female, it is believed, will be mates for life.

Although the male leaves the den and finds another one nearby while the female bears her young, the boar doesn't just vanish, as some males in other species do. It actively engages in nest building. One observer saw a boar badger gathering greenery for the nest and carrying it to the den in an unusual manner: Its hind legs rigid, it used one front leg to pull itself forward, actually making a sled of its body, carrying the fernlike bracken between front and hind legs in a neatly arranged bundle with the stalks on one end and the fronds on the other. The badger deposited it before the opening of the den, backed down and pulled the bracken in with its forepaws.

The sow badger was waiting in that three-story-deep labyrinth (which the pair had dug after mating) at least fifty yards into the hillside. Together they arranged the nest for the birth. As usual with badgers, there is nothing ordinary about it: The gestation period is seven months. Badgerman Ernest Neal explained that after fertilization the blastocyst or early-stage embryo does not become embedded in the uterine wall for about five months, during which time there is little growth. Implantation finally occurs during December or early January, and the young are born eight weeks later.

There can be as many as seven cubs, but the average is two or three, which are born blind and almost hairless, mewing like kittens. Their eyes open in ten days, but the cubs do not venture far from the den until they are eleven weeks old. They stay with their parents, learning the survival art of their kind until late September or early October. The mating instinct also stirs in them at the age of seventeen months, at which time they can reach a height of nine inches at the shoulder, be as long as thirty inches, weigh up to twenty-four pounds but probably will average fifteen, and live for ten years in the wild or longer in captivity.

Often, in August, several families will live in one large underground home while the other dwellings get an annual cleaning. The entire family does a thorough job, cleaning out all stories and side chambers, neatly spreading the debris in a fan shape before the exit. One curious naturalist hauled away five cartloads of the results of the "spring cleaning" of one badger home.

Badgers do not hibernate in the true sense of the word. Winter does dull their appetite and make them drowsy, and they do go deep into their cozy retreats when the temperature is below zero. But hunger can stir them into action at any time, and they have been seen prowling the outside world during all hours of winter.

It is believed that the badger got its name from the French *becheur*, "a digger." Without fear of anthropomorphizing, it is safe to say that the badger is much more.

·6·

RED FOX

IF THE BUILDER of our house in Connecticut could have seen what would happen in the next few minutes, he wouldn't have euphemistically called the large window giving a wide-angle view of a meadow, some shagbark bushes, oaks, maples, and beeches a "picture" window, but what it really is—a true window on the wild.

This sparkling August morning I was sitting at the typewriter struggling for the words to work from my head onto the forbiddingly blank piece of paper at which I was staring. I looked up, the peace of the quiet morning beckoning, that early sun gently glistening on dew that beaded the grass.

Suddenly a figure appeared; not much larger than a domestic cat, it trotted confidently across the meadow and sat staring directly at me.

The world stood still.

This was one of the shyest of all wild creatures, a red fox pup.

Cocking its head, it looked up at me with curiosity, innocence, and insouciance. Separated only by that magic pane of glass, it was less than twenty feet from me.

Abruptly it fell to the ground and began rolling, then started chasing its tail, puppy-fashion. It was in fine coat, reddish buff, and it played in the sun for about five minutes. Then it sat up again, resuming its intent, topaz-eyed scrutiny of me.

This was a first-time experience and I found myself trembling with suppressed excitement.

Wildlife young are the symbols of what is right with our world. They are still unspoiled by suspicion, untainted by fear. There is a bouncy curiosity about them, a friendliness for the big, unexplored world that will vanish in weeks, even days.

One can become maudlin sitting staring through a window at a young wild fox that stares back. I wondered if I was the first human that this little fellow had ever seen, and I thought about what its life ahead would be like. I knew that it would soon learn that creatures such as myself, sitting huge and motionless framed in that window, were not to be regarded openly but furtively from the shelter of forest and shadow.

That's the word for foxes—"furtive"; they must become furtive to survive. But that would be a later development for this bold youngster. I then considered rising and walking to the door to see if the fox would flee. But I didn't. This was a moment, an experience, to hold, not disturb.

The fox pup, probably recently weaned, cocked its head one way, then another, continuing its examination of me. Then, from a privet hedge at the far end of the meadow, a head cautiously poked through, and I heard three sharp, yipping barks. The young fox immediately wheeled, trotted to the hedge, and vanished. The vixen, its mother, had summoned it. Enough was enough. I have seen red and gray foxes since, all adults, but never such a sight as that pup exploring a new and curious world with all the sweet innocence of the very young. I felt as if I had won some sort of extremely valuable award and went about the rest of the day in a semidaze. For a time that young fox and I were in silent communion that wiped away the enmity of ages.

Young or adult, red foxes are fascinating creatures, among our most elegant, elusive, and enduring wild animals. In fact, the fox's intelligence is so high that even though humans have been waging war against the cunning creatures for centuries, the contest is far from over. It promises to go on forever—as long as there are people and foxes.

One type of contest, that of fox hunting from horseback, holds the fox in such high regard that the animal is never killed unless it is actually caught by the dogs. These fox hunters regard the chase itself as the sport, not catching the quarry. One of the largest of the one hundred hunt clubs registered in the United States, located in Virginia, claims that it kills less than one percent of the foxes it chases.

Other fox hunters, however, who use dogs and guns, do bag the fox, considering it a major game animal, and sell its hide when they can, but they always take great satisfaction in outwitting this animal that more often than not proves to be smarter than they are.

And there are fox hunters who chase the clever animal mainly to hear the "music" of their hounds as they trail. They do kill the fox if their dogs are good enough to bring it to bay, but the kill is not the important part of this sport.

There still are trappers who take the fox, but its fur is no longer popular and is not the prize that it once was, making the price of the skin hardly worth the effort. And there are many hunters who consider the fox an enemy, shooting it on sight if they can, believing that it kills rabbits, partridges, pheasants, and quail. Farmers, mainly those who are careless with the housing of their poultry, also kill and trap foxes. So there are many hands raised against the fox, many enemies. But there always have been.

All large carnivores are its traditional enemies, and naturalists believe that the lessons learned by generations of foxes evading these larger predators, plus the fact that the fox commonly plays the dual role of hunter and hunted, has developed its brain to a high degree.

Although they are swift and can run as fast as forty miles an hour for short distances, foxes tire rather easily and depend upon skill and complete knowledge of their range rather than attempting to outrun a dog pack.

That territory, where the fox spends its life if it is varied enough and contains sufficient food, averages about nine hundred acres. Biologists believe that foxes that range over five miles from home are out of their territory. A hard winter can make the animals wander this far or farther in search of food.

Thus, in the range where the fox knows every woodchuck den, hollow tree, cave, and stone wall, it is master of the terrain, and fox hunters have a difficult time running it to earth. These frustrated fox hunters have spread the intellectual fame of the red fox. They are all too familiar with what a superb hunter it is. It has mastered many ruses and tricks, and, most important, it realizes the value of scent, which it uses every day in securing its prey, or food.

The fox expertly uses rocky ledges and stone walls, which are silent trails and poor conveyors of scent. Nature writer Sharon Saari reports that one hunter watched a fox being trailed by dogs on snow-covered ground leap to a high stone wall that encircled a cemetery, thereby bewildering the tracking hounds, which lost both scent and ground trail.

Naturalist Archibald Rutledge saw a fox chased by hounds run halfway across a log that spanned a small stream. It halted and looked down at a small area of rocks and soil forming a tiny island in the stream, then leaped down onto it. Instead of jumping from that to the opposite shore, away from the hounds, the fox leapt back to the shore it had just left. The hounds came howling up to the stream, found the trail on the log, and crossed to the other side. There was no fox scent there. Milling around for a while, the dogs finally gave up the chase.

Several observers have reported foxes pursued by hounds running closely behind a fleeing deer, thereby confusing the scent trail. If the deer leaps over a fence, the fox drops back and follows the fence line, with the dogs usually following the deer over the fence. Other basic tricks of the fox are backtracking up streams, deliberately making a confusing crisscross of tracks, and circling and coming behind the dog pack.

Foxes have been seen twisting in and out among a herd of grazing cattle, completely confusing the scent for the pursuing enemy. Some fox hunters claim that foxes have been known to ride the backs of sheep to destroy their scent trail, and even run agilely along a high rail fence. One veteran fox hunter swore he saw a fox being chased through snow dragging its bushy tail to wipe away its tracks.

Consider also the oft-repeated, documented account of two fox hunters and their pack of skillful hounds. It didn't take long for those seasoned hounds to find fresh fox scent. The hunters went to the top of a hill to watch. Soon the dogs bayed the fox into sight, running fast; the dogs were quite a distance behind.

The fox made two wide circles, backtracked, and emerged in front of the hounds, then quickly lost them again. From their vantage point the hunters saw the fox enter a large hollow log, rest briefly, then burst out the other end. The hounds finally caught up to the scent trail where the fox had entered the log. Sniffing eagerly, the dogs picked up the trail at the exit hole in the log and bayed on.

The fox made a wider circle this time, forcing the hounds to cover more ground. Once again the fox went into the log, rested briefly, then came surging out the other end with renewed vigor.

Held inactive by this fascinating drama, the hunters watched the fox repeat this "holing up" procedure three times until the pursuers were staggering with fatigue. Baying weakly, the dogs again trotted after the fox that popped out of the log; the hunters finally went down the hill and poked a stick into the hollow log. Out leaped a fox. There had been two foxes, playing a game of relief-runner. When one fox got tired, it entered the log, and the rested animal resumed running.

The fox sometimes plays a more deadly game. Respected naturalist Ernest Thompson Seton reported an incident of hounds driving a fox from the woods near Canaan, Connecticut. With the dogs in close pursuit, the fox led them up a railroad track before an oncoming train. Timing it precisely, the fox leaped from the tracks just in time, leaving the slower dogs to be destroyed by the fast train.

Similar reputed performances, perhaps fanciful stories of the animal's remarkable intelligence, may have first started the legends and fables of the fox. Over two thousand years ago the slave Aesop, from the Greek island of Samos, wrote eleven famous "fox fables," among them "The Fox and the Grapes," "The Fox and the Cat," and "The Fox and the Crow." When these fables were written, Greece was ruled by tyrants who prohibited free speech. Aesop used his fox fables for political purposes to illustrate morals that he could not otherwise state: Do not trust those who flatter. Cunning can outwit itself. One bad deed deserves another.

The fox has also added colorful words and phrases to our vocabulary. "Outfoxed" for outsmarted or outwitted; a "foxy fellow" is sly, crafty, or clever as a fox.

If one word can describe the character of a fox, perhaps "clever" is that word. Consider how the fox uses that cleverness and ingenuity to aid its own comfort.

One man documented a report of seeing a fox carrying a piece of bark in its mouth. He watched while the fox waded into a creek, completely immersing itself, leaving only the barest portion of nostrils and the piece of bark above the water. The fox was submerged only briefly. Then it came out of the creek, dropped the bark, shook off the water, and trotted away. The observer examined the bark. It was covered with fleas.

Another naturalist saw a red fox carefully pull wool from a barbed wire fence. Then, wool trailing from its mouth, the fox backed into a stream. Another observer watched a fox peel silk from an ear of corn and wade into a river with the silk in its mouth. In both cases, when the foxes left the water, the wool and the cornsilk were full of fleas. All three actions were well documented. One naturalist wrote a long article footnoted with dates, names, and affidavits about this flea-divesting device of the fox.

That cleverness is overlaid with another characteristic which makes the red fox the unique animal that it is: courage. Not long ago a biologist reported seeing a female with three young fighting off two hounds several times her size, delaying them while her kits got back to the safety of the den. The biologist drove the dogs off, saving the vixen.

Many cases were recorded during the period when fox

trapping was popular, in which females caught by traps were known to have dragged them for days to reach the den to suckle their young. In one case, when the female was killed and the young were alive in the den, the male brought a tempting variety of small game to the starving, unweaned young, even regurgitating food for them, but to no avail. The kits died.

I am convinced that red foxes are the most devoted of all wildlife parents. I still recall with horror one experience that dramatically and tragically proved this, at least for me.

Two government predator-control agents gassed four fox kits in their den. The gas ruptured their lungs, driving them out to die. The agents buried them, leaving only their tails exposed. This was done to attract the female fox so they could trap and also kill her. Knowing, with her remarkable scenting ability, that her young were dead, nevertheless the female returned and attempted to dig them out. She didn't succeed; steel traps were set where she was certain to dig. Despite the pain and hazard of the trap sprung on her front foot, the mother kept digging until she was clubbed to death. Later the male was also trapped at the den while trying to rescue his family.

This occurred in South Dakota, where there was an all-out war against the fox using gas, fire, poison, guns, propaganda, and bounties (money paid for dead "varmints"). It was money that prompted this particular "death to the fox" campaign.

The pheasant population of South Dakota usually peaks at about 15 million, bringing out-of-state sportspeople and more than a million dollars annually to landowners and businesspeople of the area. In 1964 the pheasant population dropped to about 5 million; the income from sportingpeople sank well below half of normal. Outraged politicians and businesspeople pointed at one culprit, the fox, and, to a lesser degree, the raccoon. The Democrats, who were out of power, claimed that if they got in they would wipe out the fox. In less than a year the politicians that were in power paid just under $100,000 in bounties for 25,301 foxes, which included 12,436 pups barely out of the den.

Conditions worsened. The pheasants didn't prosper; the foxes were slaughtered. What actually caused the problem was an extended drought that not only drastically reduced the pheasant population but brought an emergency drought designation to almost 500,000 acres of soil-bank land (the best pheasant cover), which were then grazed and mowed by the state in less than two weeks. This happened at the peak of the pheasant nesting cycle. Finally the governor of South Dakota acted, ordering a study by joint committees of the two state universities.

The study bluntly stated that habitat was the most important factor in keeping a high pheasant population; that the birds needed sufficient shelter, food, and breeding grounds. "The fact first of establishment," the scientists pointed out, "then the depletion of soil bank acres was profound." That waste of habitat, and that alone, destroyed the pheasants.

The scientists also noted that predators were natural components of the pheasant environment, and their study proved that the large reduction in the number of foxes resulted in no change at all in pheasant populations but in a sharp and dangerous rise in the rabbit and rodent populations.

This state's action is mentioned in some detail because in essence it is the story of what has happened to the fox. In state after state almost the same scene has unfolded: The fox was slaughtered; the state reconsidered and asked for scientific help, which time after time proved that the fox was not the villain, we were. Yet the bounty system still continues, wasting millions of dollars in some states.

After nineteen years of bounty hunting for foxes in Michigan, at a cost of 2½ million dollars, there remained 75 percent more foxes than when the bounty was initiated. Normal licensed and controlled sport hunting, some trapping, and natural mortality proved a far superior substitute to regulate numbers.

Actually, the red fox is far from a wildlife villain, but neither is it a wide-eyed innocent whose every act benefits us. This is simply one of the brightest, most interesting members of our assembly of wild animals that skillfully does what nature intended it to do.

It was believed that English colonists brought this wildlife asset to America. Unhappy with hunting the tree-climbing wild gray fox that dwelt mainly in forests, the fox-hunting plantation owners of the South did import red foxes from England, but there always was a red fox in the United States. In Idaho red-fox fossils have been found that date as far back as fourteen thousand years. Today it exists in twelve subspecies throughout most of the North American continent, from Nova Scotia, Labrador, and Newfoundland through Minnesota, Utah, Mexico, California, and Washington, to British Columbia and Alaska. These subspecies, according to location, vary in size and in color.

Closely related to the dog, coyote, wolf, jackal, and others of that family *Canidae,* foxes belong to a different genus, *Vulpes,* which are smaller, more agile animals with sharper muzzles, larger ears, and longer, bushier tails. The red fox has forty-two teeth, the same number as the dog. It also uses barking, yelping, growling, sharp yips, and even screams as means of communication. Although they average about forty inches in length, twelve of those are their bushy tail, making them look larger than they are. Surprisingly they only average about ten pounds in weight, although some males have topped fourteen pounds. Their life span averages about eight years.

The fox that most of us see is *Vulpes fulva,* from the Latin "yellow fox." Actually it isn't yellow; the coat looks more like burnished gold. (That handsome coat is changed every year, the thicker, new coat becoming fully furred by late December and the tail fur changing in late spring.) The color may run from a deep rust to a pale gold that is accentuated on the back and behind the head, running to a lighter shade as it nears the belly, which is white. The tail also is a deep rust or gold, flecked with black hairs; the four-inch tip is pure white. The insides of the ears, throat, cheeks, and chest are also white. The nose, rear portion of the ears, and lower legs are black. The red fox's eyes are topaz, a bright yellow. Its five rows of long whiskers are also black.

The red fox has three rather rare color phases, which sometimes occur. These vary from the normal golden-red: silver, almost black-coated with white guard hairs; deep black with white tipping the tail; and the striking "cross" fox, brown-yellow with black underfur and with black striping back and shoulders. Occasionally all four colors will appear in one litter. The red fox's natural range is all of the United States except the southern Atlantic coastal area, lower Mississippi, southwestern deserts, Pacific coast, and south-central Great Plains.

Of the five species commonly called foxes, only the red, the kit, and the swift (rare, small desert foxes) are true foxes. The gray and the arctic foxes belong to a different genus and are not members of the *Vulpes* family.

Both sexes mature at ten months, but the female sometimes will not mate at all during her first year. She is monoestrus, having only one heat or estrus period a year, which lasts for two weeks. But she will only accept a male during the middle of that period. The male, however, is ready from late November right through April, and roves widely seeking a mate. In heat and unmated, a female may even range a wider area than the male to find a suitable mate.

Males will fight fiercely, but not to the death, over a female they both have chosen. So concentrated and fierce was one such fight that two foxes fell over a cliff and were killed.

Fox expert Leonard Lee Rue III says that foxes use their bushy tails when fighting. As each circles, it carries its tail curved and flips it in the other's face, trying to blind the other momentarily in order to dart in and make the sharp bite that may send the bitten one scurrying in defeat. They also push each other off balance with their hind quarters, standing upright and throwing their paws at one another.

When the victor mates with the female during a four-day period, several fifteen-minute couplings may occur. During this period the pair are playful and inseparable, often mock-stalking or wrestling with one another and making kitten-coy approaches. The courtship itself is long, lasting about three months, during which the pair hunt together. Their team efforts are most effective; sometimes one fox drives game toward another which lays hidden.

Mating usually occurs in February. It is believed that they pair for life, although some naturalists dispute this. During the approximately fifty-day gestation period, they both investigate dens in their home territory of one or two square miles. Often abandoned woodchuck dens are selected. Three additional dens are cleaned out and held in readiness should an emergency arise at the home den and the young need to be moved. The den is thoroughly cleaned, and is often enlarged to the point that a mound of earth two feet high and four feet in diameter is dug out. Dens can also be caves or hollow trees. Seldom do the red foxes excavate their own.

The male does not stay in the den with the female and the young after they are born. The female does not leave at all for the first ten days. She has one litter a year, averaging five kits but running from three to ten. The young, which are born blind, weigh four ounces and are well coated with sooty brown fur. Their eyes open in about ten days, but it is a month before the playful pups appear at the den entrance. During the first two weeks after the young are born, the male brings food to the den for his mate.

Lactation lasts about two months. During this period the female (still with help from her mate) supplements the offsprings' diet of milk (which is three times richer than cow's milk) with half-digested flesh which she regurgitates, tempting the pups to eat and stop nursing. After lactation she continues to feed them, first on dead small mammals, then live ones, which the pair both provide.

During all this time the male remains near the den, acting as a protector and provider, bringing everything from moles to muskrats to his family.

When they leave the den, usually only when they are called by their mother, the pups are playful and tend to be independent. But they are carefully watched by both parents, who control them by the sharp yip summons I heard, as well as physically. One observer watched a female with a single pup use her front feet to push it down when it refused to remain still when humans were passing on a road a hundred yards away. If, in playing with their parents, they nip too hard, the pups are stopped with a low growl and a rough push. But they are pampered, being fed about ten times a day by the female.

It is the vixen that gives the first hunting lessons, when the young begin to leave the den from age five to ten weeks. The pups start on grasshoppers; then, when they are fast enough, they graduate to mice. The father helps then, showing them how to stalk rabbits and woodchucks. One observer saw a big male bring a cottontail back to a den and then, with four young watching, carefully peel off its skin. They are taught to use their eyes, ears, and nose, and to swim.

Some prey is taken by running it down, some by masterful stalking; others by the fox remaining motionless until the prey approaches. Then it is caught, with the fox making an agile, catlike leap.

The parent foxes often move carefully with their young through wild meadows during early morning or at dusk, hunting for mice. They move so gracefully that they almost seem to be on tiptoe like ballet dancers; this impression is strengthened by their black lower legs and feet. They teach the young to do this meadow-stalk after mice, which consists of careful, slow movement, much work with the nose, eyes, and ears, then the leap, made with impressive precision and grace. The quick red fox is such a master that rarely does it miss a mouse. The young do, though, often, until they get the hang of it. Long after most wildlife young are on their own, the fox family is still together, with the parents passing on survival secrets that have kept this remarkable animal with us despite every effort to wipe it out.

These lessons in everything from the art of backtracking to

the value of complete mobility and using terrain as camouflage go on until the family breaks up in the fall. Until then a strong social bond exists that not only keeps them together, but unites them in protecting their area from other foxes. Hidden caches of food, the dens, the resting areas, the hunting territory which may consist of anything from a large meadow to a couple of square miles, are "marked" by scent—male urine—telling others of their own kind that they are off limits.

If a fox is providing for a family and there is a careless farmer nearby who thinks so little of his poultry that he doesn't fence it or keep the henhouse locked, then there are going to be missing chickens. But mainly a fox doesn't need to steal. Besides being exceptionally smart, its speed, swimming ability (it can swim five miles an hour), radar-keen senses, and, more important, its omnivorous appetite (the fox eats everything from wild berries and grapes to snakes, tadpoles, frogs, insects, and rodents) make all outdoors a pantry. Its favorite food is mice, then cottontail rabbits and squirrels. Some birds also are taken, but it is mainly the weak or ill; thus the fox benefits that species.

It also is important to remember that without one of nature's levelers such as the fox, a single pair of grouse would in less than six years pour into our woodlands 33,000 young, which would in their excessive numbers bring them starvation, disease, and epidemic.

Paul M. Kelsey, conservation educator of New York State, places the feeding habits of the red fox in intelligent perspective. "Traditionally the fox is blamed for poor hunting," he says, "but foxes, being skillful and clever, are predator-opportunists, always taking food that is the easiest to obtain. Most of their protein, therefore, comes from mice, the most abundant food. In summer and early fall, fruit is a major food."

Many studies have been made of the fox's eating habits that strengthen this thesis that it is our benefactor. Dr. John Wood of Jacksonville University discovered that rabbits, rodents, and small birds make up about half of the fox's diet. The remainder consists of harmful insects, wild fruits and seeds, even carrion.

Another scientist who studied foxes closely for thirty-five years found through careful examination of dens and droppings that their main food was mice: meadow, harvest and white-footed. There are thirteen hundred types of these rodents in the United States that destroy almost everything that isn't metallic, killing wild and domestic birds, causing almost one billion dollars' worth of damage yearly to crops, grains, and supplies.

Paul L. Errington, research zoologist at Iowa State College of Agriculture, perhaps best summed up the naturalists' reaction to the injustices we needlessly inflict upon beneficial and rewarding wild creatures such as foxes. "It is unfortunate," he said, "that man, the specialist in evil, sees in predation among wild animals so much evil that isn't there."

·7·

GRAY SQUIRREL

EVERY TIME I WATCH a gray squirrel move swiftly through the treetops with such skill that it makes me hold my breath, I wonder if, long ago, acrobats and high-wire performers got the original idea for their acts from watching these unique animals move through tall trees with the greatest of ease.

This 26-ounce acrobat is America's best-known wild animal, and probably our favorite. I recently read that this entertaining little creature that we watch and feed with much pleasure in our parks is also most intelligent. I was aware of this, but it is encouraging to have one's beliefs strengthened.

Consider the observations of two squirrel watchers who reported their findings to New York State's *The Conservationist.* Charles Chappell, of Syracuse, New York, wrote:

> This spring as I was walking along a very busy street, I noticed a gray squirrel dash out into the street, and then back to the safety of the curb, as he was nearly hit by a passing car. Before I had passed by, he had made another foray into the street and again had to scamper back to safety.

have its resident grays whose agile performance delights everyone. Cheered by their bounce, fascinated by their acrobatics, susceptible to their saucy begging, people sit by the hour all over America watching the squirrel show, feeding them nuts, photographing them, tempting the animals to climb on their arms and perch on their shoulders.

I saw an elderly gentleman in elegant attire spend the greater part of an hour in New York City's Central Park convincing a haughty old gray to come sit on his arm. In his fascination with the game, the man dropped all dignity, crouching and kneeling, making soft purring noises until the squirrel finally seemed to humor him and scampered up on his arm.

There is a distinct difference between park squirrels and wild squirrels; the former are accustomed to humans and seem to take them in their stride. Squirrels that live in their natural environment have learned that it is wise to shy away from people. And they are adept at it.

If you've walked through the woods where there are gray squirrels, you know that they are masters of camouflage, freezing to a tree trunk or lying along a branch until you have passed. One of their cleverest tricks is "sidling," being where you aren't by rotating to the opposite side of a tree. I once walked slowly around an oak, fifty yards from it, pacing that nimble maneuver. Moving as if on a greased pole, the squirrel twirled to the opposite side twenty times before it tired of the game and scampered to the treetop. The agility of squirrels in trees is astounding, permitting them to leap fifty feet—and, surprisingly, they are fast on the ground. A British biologist clocked one running beside his automobile at 18 miles per hour.

Whether in the wild or near humans, the gray squirrel is an

interesting, appealing animal that has tempted many to try to tame it. One naturalist succeeded so well that his house-broken squirrel sits at the dinner table with his family.

I am totally against trying to make pets of wild animals, as it usually ends in tragedy—for the animal. But from the evidence to the contrary in the case of the gray squirrel, I must admit that it sometimes can be done . . . But cautiously: The experts claim that all gray squirrels, tame or wild, have a habit of biting, sometimes badly, slashing fingers to the bone.

The squirrel is a fascinating creature that we must continue to remind ourselves is wild. Several decades ago opera singer Florence Hinton, presented with two tiny grays by a boy who wanted to hear her sing but had no money, became so fond of the squirrels that she finally acquired six. Against the advice of experts who said it couldn't be done, she began trying to train them to perform. It took three years to teach them to play dead, turn somersaults on command, swing on flying trapezes, take people-baths in tiny tubs, and walk tightropes. The only act of its kind, it became so popular that Mrs. Hinton gave up singing and trouped her squirrels.

"They were terrific troupers," she says. "Each performance improved, and you could tell that they really appreciated audiences and played up to them."

I haven't found gray squirrels that cooperative. They invade my bird feeder and drive away the songbirds; they try to get into the house on cold winter days—and have succeeded. I do enjoy watching them from my window, and I do appreciate the way they successfully handle life when many wild creatures are endangered.

Apparently President Eisenhower shared my feelings when some gray squirrels that were causing him annoyance created a national incident. It seems that the squirrels were eating the grass on President Eisenhower's White House putting green.

The annoyed President sent out a call for help. An electronics expert came and made high-pitched noises that aggravated everyone around the White House but the squirrels. Then the United States Army Signal Corps arrived with more electronic sound that the gray squirrels completely ignored. With his putting green in a mess, the President finally sent for a trapper, who was successful. But the gray squirrels had friends in high places: A senator from the West aroused his public with the cry, "Save our squirrels!" and started a fund to save them from this cruel banishment. Result: They were readmitted to the White House lawn.

Even the best scientific brains of the Bell Telephone laboratories discovered that the gray squirrel was too much for them. In Wisconsin squirrels were chewing up about a half-million dollars' worth of telephone cable a year, eating the lead sheath in which telephone wires were wrapped. When they gnawed it away, moisture penetrated the paper insulation, short-circuiting the line.

The scientists approached the problem as one of appetite, believing that the squirrels were attracted to the cables because they lacked calcium and phosphorus in their diet. Salt discs were tacked to the poles. The grays tried them, then went back to the cables. Before they gave up, the Bell engineers tried over a hundred possible solutions—electronic shock, iron barriers, paint, rodent repellent, weasel scent, even a steel tape armor. But the gray squirrel, as is its custom, ate what it wanted when it wanted.

This may be the key to squirrel survival in a world that is rapidly losing its wilderness and with it the natural habitat of forest creatures. But civilization's march doesn't deter the gray. It has perhaps the widest range and highest degree of adaptability of any animal and can take care of itself in any terrain, weather, or food situation. Forest, mountain, plain, public park—it's all the same to the gray squirrel. Adjusting diet to the season, their menu runs from bark and pine seeds, fungi and tree buds, through all the wild fruits and berries. I have seen them in a bee tree after honey, and have even watched them gobbling the larvae of Japanese beetles and gypsy moths.

They tackle just about anything. I surprised a gray squirrel rolling a huge egg from the nest of my white Emden goose, moving it along as agilely as a basketball player. Grays regularly eat the fish placed outside for our cat, and now they have taken to the dry cat food. (It took us a number of weeks to convince our Siamese that these hard little pellets were actually food. The squirrels knew it instantly.)

As everybody knows, their favorite food is nuts—hickory, acorns, beech, butternuts. Much misinformation has circulated regarding the squirrel and its preferred food. They do not store nuts for the long winter ahead in one central cache. They bury them in many places just a few inches underground.

Sitting quietly, I have watched one gray squirrel bury fifteen acorns in as many places in less than 10 minutes. A more patient man than I, naturalist Ernest Thompson Seton, clocked the gray as burying five nuts every 3½ minutes. He claimed that in three months the normal gray squirrel will bury ten thousand nuts.

At University Farm in Mitchell, Indiana, Walter Hahn worked out a mathematical analysis of this amazing nut-gathering ability. Estimating that each white oak on the farm's 180 acres bore from 2,000 to 8,000 acorns, Hahn observed that, before November first, this enormous crop had been so efficiently harvested that not a single nut remained.

He estimated the number of nuts at 100,000 per acre and computed that in two months less than a hundred squirrels had buried 20 million acorns.

Contrary to general belief, this industrious nut gathering is not indicative that a hard winter lies ahead. It is just the result of food abundance, which the squirrel takes advantage of. There has also been much discussion as to how the squirrel locates these many nut caches. Most naturalists agree that it is done by a combination of memory and superior scent, with scent the most important.

One scientist, Dr. Robert T. Morris, experimenting with this scent factor, found that grays went immediately to acorns buried one inch deep. The same squirrels took a bit longer to find nuts buried three inches deep. Yet three acorns accidentally covered with six inches of soil were found by one of the test grays in a matter of seconds, which seems to indicate that sense of smell varies in individuals.

Another experimenter watched a gray squirrel run across two feet of snow, stop and burrow, and come up with an acorn.

I found that the gray also has scent selectivity. Watching for two hours while a big gray buck sat beneath an oak eating acorns, I saw the squirrel hold each nut with its forepaws. It first sniffed, then sometimes discarded a nut. Later, when I examined the nuts that were left, I discovered that every one was wormy.

Naturalists contend that we owe the gray squirrel an eternal debt of gratitude. Walnuts, butternuts, and hickories won't sprout unless they are planted. Those that the squirrels bury and do not find grow into trees; thus without the gray's activity, many of our wild nut groves wouldn't exist.

I first became interested in this astonishing gray animal in the mid-thirties in western New York State, when I thought I was witnessing a gray squirrel invasion. It was an amazing, even a frightening sight. Thousands of the animals came out of the woods and scampered across roads, along streets, actually moving through villages and towns in a mass movement that had everybody baffled. Many were drowned in rivers, killed on highways, and destroyed by confused people, but they continued to move en masse for several days. Theories were projected at that time, but I know now that all of them were wrong.

Since then I have noted that many of these mysterious movements of grays have occurred throughout the country. One, read into the *Congressional Record* by Robert Kennicott, noted that a half-billion gray squirrels crossed Wisconsin in a movement that took over a month, a migration that was repeated at five-year intervals for some years.

Naturalist W. J. Hamilton, Jr., claiming that only gray squirrels make this astounding mass movement, reported that he has seen thousands swimming the Connecticut River between Hartford and Essex. He believes that it is emigration (not migration), the complete abandonment of the home region, and attributes it to three drives; search for a better climatic environment, lack of sufficient food, or overpopulation.

In fact, he studied the same emigration I had seen and came to the conclusion that it was caused by overpopulation. The two winters before my squirrels left had been mild, with plenty of food. This resulted in litters of four to seven, rather than the usual two or three young. "The population density became so great," Dr. Hamilton concluded, "that emigration was imperative for the good of the race."

This would be typical of gray squirrels. They always do what is best for themselves. Nature, of course, has aided in this talent for self-preservation. All the squirrel's senses are excellent. It is one of the few animals that has perfect, uninterrupted vision when it looks upward. The retina of the gray's eyes is made up only of cones which pinpoint perception of movement to a remarkable degree. If you've been wondering why there aren't more accomplished squirrel hunters, this may be the reason. Few of us have the ability to freeze into immobility for long enough periods to tempt these X-ray-eyed animals into the open.

Masters not only of motion, grays are also creatively adept when it comes to camouflage. They can look so much like part of a tree that they even fool foresters.

That bushy tail may be the most useful asset of this versatile mammal which belongs to a family with 1,300 varieties of squirrel. All the five species of grays that have been fascinating us since the *Mayflower* docked use the tail advantageously. One winter I saw a gray teetering in the top of an ice-sleeted oak. It lost its balance on the slippery surface, falling at least seventy-five feet to the frozen ground. Unfurling that tail like an umbrella, it landed apparently unhurt, flicked its tail, and scampered off.

Another observer watched a gray jump off a cliff, undoubtedly unaware that there was such a drop. It fell six hundred feet, fluttering rather than falling, actually swimming its way down, using all four legs as if it were in water and spreading its tail like a parachute. The naturalist, estimating that it would take any other animal twelve seconds to drop that distance, said that it took the squirrel forty seconds. He said it landed almost as easily as a bird and was unharmed by the long fall.

That 8½-inch tail can also be draped over its head to

protect it from rain and snow, can be wrapped around it for warmth in the nest, and can be used to maintain that exceptional balance as it moves from branch to branch and tree to tree. I have seen a gray squirrel with half its tail gone fall three times as it was chased through a grove of trees, which is dramatic proof that the tail is not the least of its appendages when it comes to survival.

Our Indians called the gray "He who sits in the shadow of his own tail," and the Greeks, who had a talent for naming everything from an olive to a god, called him *skia oura*, "shade tail," which yields its scientific name, *Sciuridae carolinensis*.

These playboys of the treetops don't just run away to play another day. If they are faced with danger they know how to handle it. J. F. Bogen of Centralia, Illinois, saw two grays battle a huge black snake. The snake quickly coiled around one. Death seemed certain until the other squirrel flashed in, grabbed the snake's head, and hung on until the squirrel was released and the snake was dead.

I have seen a gray pop out of a nest evidently full of young, stand upright and box so furiously at a red-tailed hawk that the bird finally gave up and went after less aggressive prey. When the bird was well away, the squirrel went into the nest and reappeared with three youngsters clinging to her back, their tails and legs wrapped around the mother's neck. Taking no chances on the reappearance of the hawk, she was moving her quarters immediately.

The young are born in a fall or spring litter, usually averaging three to a litter. The mating of gray squirrels may be the most active of any squirrels, with several males noisily chasing a female through the treetops. When the most agile reaches the female, he takes possession and fiercely fights off the other males. Six weeks later the naked half-ounce young are born in a tidy and weatherproof spherical home that is interwoven with leaves and twigs, roofed with grass thatch, and has two camouflaged entrances. This nest in the top of a tall tree is also insulated inside with grass; any openings are filled with moss and shredded bark. There is usually another larger, airier home in the vicinity where the family moves after the young have been suckled for seven weeks in the nursery nest.

Gray squirrels do not hibernate, although on exceptionally cold days they will not venture out, but will sleep through the harsh spell in a hollow tree that they have lined with grass, leaves, and often even a cache of nuts.

The gray squirrel may indeed be our most entertaining and widely watched wild animal. Devoted squirrel-watchers, however, believe that it is always at its best in its natural world. I have watched a mother squirrel with two half-grown young trying to keep pace with her as she sailed through the trees. One fell, instinctively unfurled its tail, landed unhurt, scrambled up the tree, and continued its aerial follow-the-leader.

One woman was enchanted watching four young squirrels regularly meet a cottontail rabbit at the edge of a meadow at sundown for a romp and a game of tag.

Life, however, is not all acorns and play for the gray squirrel. It has many enemies; the next animal in this collection is probably the most deadly.

·8·

WEASEL

SOME NATURALISTS call the weasel the little assassin. And this animal is a professional killer—probably the most professional—which, in the natural world, means skilled.

An action of a weasel observed by my wife doesn't bear this out but does point up the fact that wild creatures, like people, are individuals with their own abilities and failings. However, the sight that my wife saw through the window was an exciting one that she claimed made any television drama pale by comparison.

Perhaps I should describe our window on the wild, through which we have seen so many of these fascinating sights. Let's take the house first. A renovated farmhouse, it is clad with shingles that have been weathered silver. It is set one-quarter of a mile back off a secondary road, three miles from a Connecticut village.

I won't detail the entire house; just the living room, where the window dominates. It's a 24-foot by 35-foot paneled room; the window is 4 feet by 8 feet, a solid sheet of thermal-pane glass facing the west. Directly before it is a 100-foot-square meadow, bordered by a variety of bushes and trees, with a privet hedge at the far side setting it off from another, larger meadow beyond. The house itself is surrounded by forest, with no neighbors to disturb that forest for about a half-mile on any perimeter.

My wife, Maria Luisa, the other half of this team, has her worktable and typewriter before that window, giving her a direct view of the meadow and hills beyond, which are always veiled with purple. That meadow is like a stage across which wild actors are often passing and performing, as was that star, the fox pup.

This April day after the last snow had melted, I was away from home and my wife was busily typing when she looked up and saw the weasel. Here is her note to me of what she observed.

> Saw weasel leave house under step at door of terrace, go its route across the terrace and yard to stone wall (around the hidden garden) and disappear. About a half-hour later it returned by same route with something very dark in its mouth. When it got onto the terrace heading for the step, I lost sight of it. Then suddenly it dashed across the terrace empty-mouthed and ran up the maple tree there. Then I saw a chipmunk in the tree (it must have been the chipmunk in its mouth, but why hadn't it killed the chipmunk?) and the weasel chased it. They ran around and up and down the tree. The chipmunk managed to elude the weasel and scurried out of the tree to the stone wall (in front of my window) and disappeared down a hole about six feet from my window. Weasel came down the tree, ran to the terrace, sniffed and ran all around, then hunted the length of the wall, stopping right in front of me where I could really see him. Then he started quartering the grass, came to the hole that the chipmunk had gone down, hesitated an

instant. Then he started down it; first his head, then his middle, then his wider hips which I didn't think could make it, and, for an instant, just the tail stuck out. After a second or two it disappeared. Weasel was in hole approximately 30 to 45 seconds, then the tail reappeared, then the hips, middle and shoulders. I held my breath, thinking it would have the chipmunk in its mouth, but it had nothing. It looked around a bit, then went to the hole under the terrace step. He was not furtive, but quick, graceful and persistent.

My wife was so excited by this unusual observation of a weasel being unsuccessful at its task that she recited it to me in detail three times during the following day.

There was a surprising and beneficial aftermath to having a resident weasel. As all country homes do, ours had a large tribe of mice that inhabited our cellar. Because they are very destructive, we tried to keep them under control with a rodent killer mixed with ground grain, which we placed in cardboard trays.

Shortly after Maria Luisa saw the weasel, I went to the cellar to refill the trays. They hadn't been touched. This had never happened before. Then I remembered. We had a resident weasel, perhaps the most effective of all rodent killers. Had its skill driven the mice out of the cellar? Could it have killed *all* the mice? Anyway, we had a mouse-free cellar for three months. What happened then I have no way of knowing. Did the weasel move into a new territory? Was it a female with a family that took them away for any number of reasons?

Another question bothered me: Why had that weasel failed to capture the chipmunk? It is rare that a weasel bungles a job of killing. If it was so good as a mouser, what happened in the chipmunk episode? Had its rich pantry of mice satisfied its appetite and slowed it down? Did that chipmunk have a smaller escape hole through which even the sinuous weasel couldn't follow? Why did that skillful animal lose the chipmunk in the first place? Practicing? Was it a young weasel that had not yet perfected its skills? Was it just playing cat-and-mouse with the chipmunk?

Speaking of cat-and-mouse, I saw a drama of cat-and-weasel that I won't ever forget. One summer afternoon I had gone to a nearby river to fish, hoping to get a trout or two for supper.

Deciding the fish had won the contest, I waded out of the stream, laid my rod on the ground, tucked my back against the elephant hide of a beech trunk, and watched the woodland world go by. For a half-hour there wasn't much: two gray squirrels; a chipmunk; five noisy, sparkling bluejays; a painted turtle, its carapace edged in crimson, moving as if it were made of solid lead, its head going from side to side like a snake's. Then the curtain went up.

I don't know how long either of them had been there. But suddenly there they were accosting one another on the narrow trail that ran along the river—a long, skinny, gray woods cat (a domestic feline gone wild) and a long-tailed weasel. The cat acted first, pouncing immediately, catching the weasel on its hindquarters as it twisted to escape. In the battle that followed the cat reminded me somewhat of the man who climbed a tree to catch a bobcat and shouted to his friend below to come up and help him let go. Although it was an uneven contest from the outset, since the cat was several times the size of the weasel, the smaller animal acquitted itself nobly, nipping the cat, trying to get to its head for the famous death bite. Failing, it clung to the cat's right front leg. While it fought, the weasel uttered high, maniacal cries. The cat spit and growled.

In less than five minutes it was over. But the cat would remember that encounter for a long time. It limped off on three legs, leaving the weasel on the trail. I wondered why and soon discovered the reason: During the fight the weasel had emitted scent from its anal glands that was so strong that it almost nauseated me. It was nearly as strong as skunk scent. All the large weasel tribe, which includes the badger, ferret, fisher, marten, mink, otter, skunk, and wolverine—the *mustelidae* family—emit that penetrating, musky scent in varying degrees.

That cat could have been battling for its life, not for a meal, for the weasel seldom backs down from any fight. Weasels have been known to charge a man who stood in its way and have on a number of occasions even attacked dogs.

Since that day at the trout stream, I have become fascinated with the weasel, which is probably the most perfect killing machine ever devised by paradoxical nature. They are so swift and sinuous that more than one naturalist has described them as resembling a snake with fur. Also, I had been told that the weasel lived solely by killing rabbits and game birds, and I had heard of one weasel wiping out an entire flock of chickens in a single blood orgy.

In my effort to learn all I could about the weasel, I found that it is indeed a killer—a highly skilled, even frightening one—but usually, like all carnivores, it kills only in order to live, and most of its activities benefit us. In fact, Pennsylvania paid one and a half million dollars in bounties for weasels from 1915 to 1954, when people there belatedly discovered that the animals were good to have around and lifted the bounty.

A Michigan biologist making a wildlife study noted that

four resident weasels did not kill any of the ten rabbits or several pheasants that lived in their territory, but preyed exclusively on the smaller animals—mice, shrews, moles, chipmunks, and ground squirrels. Naturalists Polderboer, Kuhn, and Hendrickson made a winter and spring study on 160 acres, reporting that a population of 14 pheasants, 24 quail, and 35 cottontails were unmolested by weasels, with the exception of two cottontails that were killed by an old male weasel. Examination of weasels' nests revealed food catches of 30 mice, 4 shrews, 2 pocket gophers, 2 deer mice, 1 least weasel, and 1 tree sparrow. This study proved that weasels do not kill from blood lust as many believe, but that, like squirrels, they store food for the lean days.

Naturalist Robert Kennicott wrote of finding a cache of 100 rats and mice that were caught and hidden by a single weasel in three nights of hunting. Whether this proficient animal was planning a vacation from killing and was going to live on the cached proceeds, or whether its desire was bigger than its appetite, the naturalist didn't say. But he did remark on the fact that all the weasel's victims were problems for us.

Another scientific study at Giant Forest, California, involved a female weasel with six young. After the birth of the litter, it was observed that the weasel stayed close to her nest for thirty-seven days, rarely ranging more than 1,000 feet from it. During that time the biologists counted the kills she made for her hungry brood: 78 mice, 27 gophers, 2 moles, 34 chipmunks, 3 woods rats, 4 ground squirrels. One hundred and forty-eight kills in thirty-seven days. Not a game animal or bird was taken.

What about chickens? Do weasels kill as many as they are accused of? The scientific view is that in nine years out of ten the weasel will find sufficient food in the fields and woods, but in the tenth year there may be a scarcity and it will temporarily turn to domestic animals. It is at this time that the weasel is seen and its evil deeds noted. Actually it has been proven that rats do most of the chicken killing that weasels are blamed for.

Biologists Criddle and Criddle recorded that in a twenty-year study of weasels, with the animals allowed unmolested run of farm and poultry buildings, only *six* chickens were taken. The number of rats and mice killed was impressive. The mature weasels, which store almost no body fat, ate one-third of their weight in vermin daily; the growing youngsters ate over 40 percent of their own weight every day.

These youngsters are among the best fed, trained, and cared for in the wild family of animals. I once uncovered a nest of long-tailed weasels under an old woodpile. The wood was unfit for burning, and I was hauling it away in a wheelbarrow. The weasel family—the mother and four young—were in a soft nest of feathers and fur under the last four logs. Without hesitation the mother leaped for me, fastening her teeth in the sole of my leather boot. I shook her off. She charged again. I shook her off. Back she came again, with the young weasels, unafraid, craning their necks to watch her action. That fearless animal attacked me (at least 50 times her size) six times before I decided that she had earned my respect, and I left her. Watching from a distance, I saw her pick up each youngster by the nape of the neck and take it to the nearby woods.

That logpile wasn't the ordinary weasel lodging. Usually they prefer hollow trees or the abandoned dens of other animals. With both parents working, the nest is lined with the fur and feathers of their victims. The male often stays in the vicinity, standing guard and helping provide food until the young are at least half-grown. The family that I disturbed either had lost its father or he was away hunting.

Weasels don't do anything in the ordinary way. Even in the sexual act they are mighty—it often continues without a break for three hours. The result of that marathon wooing is a litter of one to twelve, with an average of six. In one study with the breeding occurring in August, the young were born the following May. In eighteen carefully observed cases the embryos were implanted only 21 to 28 days before the young were born: In the eighteen females studied, gestation varied from 220 to 337 days, with an average of 279 days. In the early part of the long gestation period, the embryos lie dormant in the uterus.

At one day of age the pink, hairless young of the long-tailed weasel weighs 3.1 grams, which is about 3 percent of the mother's weight (the male is often about twice the size of the female). In two weeks they grow a filmy veil of white hair; at twenty-one days they are ready to chew a little meat; at three and one-half weeks they'll take some meat solids every day. By then canine teeth have appeared in upper and lower jaws, and the incisors are beginning to break through. Their eyes open on the thirty-seventh day; in ten weeks they are ready to take over their world.

They seem to be born with what is probably the most effective killing technique in the wild, as weasels that are born and raised in cages have proved to be just as fast and as deadly as the wild ones. But females have been observed killing mice and rats with young weasels in attendance watching. Most naturalists agree that in relation to its weight the effectiveness with which the weasel kills is unsurpassed. A weasel will often take animals much larger than itself without trouble.

Like a coonhound, it hunts by scent, as my wife observed.

Its favorite killing method was illustrated for me by a weasel taking a gray squirrel. I saw the action one fall as I sat with my back against a tree, soaking in the silence of the woodland. The weasel had evidently trailed the squirrel to the oak tree, where I saw them. The squirrel went up the tree, the weasel after it. Although the weasel came across the ground as quickly as the squirrel, probably even faster, in two-foot humpbacked leaps, it was slower in the tree. Then the squirrel made its error. It left the oak in a frantic leap. The weasel didn't hesitate. It sailed from the tree, hitting the ground with a heavy thump, then streaked after the squirrel. In seconds it was atop its back. Using its hind feet to manipulate the squirrel into proper position, the weasel bit it at the base of the skull. The gray squirrel was dead instantly. The weasel then did another amazing thing. It picked up the squirrel and carried it into the woods, seemingly without effort. It must have weighed twice as much as the weasel.

That weasel was the long-tailed, the largest of the three species inhabiting North America. It was 1½ feet in length, one-third of that being the tail, but its slender body was less than 1½ inches in diameter. It is this slim, muscular body that enables the weasel to go through any hole or any burrow. Its rule of measure seems to be that anyplace it can thrust its head, it can take its body, and it can even follow a field mouse to the end of its burrow.

The ermine is the most famous of our three species, probably because the delicate white fur is so well known. In ancient times only royalty wore the fur, as peasants were forbidden to wear that white badge of honor. As late as 1937, ermine pelts were sewn onto coronation robes and cloaks; over fifty thousand were used for the coronation of England's King George VI.

The third, the least weasel, except for the shrew, is the

smallest beast of prey in the world. A full-grown male seldom weighs over two ounces. It is also the best mouser in existence, nature itself having invented what we have been striving for centuries to perfect—a better mousetrap. The least weasel is so small and so strong and fierce that no mouse can evade it. Durward Allen, the eminent biologist, wrote of a naturalist raising four least weasels in a cage. In fifty-two days they ate 252 mice; their speed in killing increased from 2 minutes to 30 seconds. Allen also wrote that in five food studies of the long-tailed weasel it was found that from 70 to 93 percent of their food was mice, rats, pocket gophers, shrews, voles, and ground squirrels. Another naturalist wrote of a pair of these tiny weasels killing a dozen mice apiece every day, eating their fill and caching the rest.

Enemies are hawks, owls, foxes, and their bigger relatives, minks, skunks, and fishers. But sometimes their predators find themselves in trouble. One hunter saw a hawk take something alive from the ground, fly aloft with it, then suddenly falter in the air and begin to fly erratically, finally plummeting to the ground. Detaching itself from the hawk's neck was a weasel. It scampered off unhurt.

A friend saw a long-tailed weasel combat a four-foot king snake, killing the powerful reptile in less than three minutes. Ernest Thompson Seton watched a cowboy turn over a saddle and have a weasel flash out and attack him, trying four times to reach his throat before the camp dog killed it.

Some scientists claim that the weasel is ill-tempered, mean. If so, biologist Durward Allen feels they have good reason to be. He believes that the animals suffer from perpetual headaches caused by a sinuslike infection called *skrjabingyliasis*. In confirmation, in an examination by other biologists, sixteen out of twenty-nine weasels were found to have a parasitical infestation of the frontal sinuses.

Weasels are not abundant anywhere. Researchers believe that we may have one weasel in residence for every fifteen acres of wild land. They also found that 90 percent of hunters, fishermen, farmers, bird-watchers, campers, hikers, and all those who get off the beaten track will try to kill a weasel whenever they can. This is unfortunate. Killing one weasel is tantamount to turning loose a horde of rats and field mice to do their damage.

It is a mighty tribe: The foot-long ermine has twenty subspecies in North America and sixteen in Europe. The common long-tailed has forty-two relatives in the United States, Central, and South America. The least is the rarest, but all three species are seen nearly everywhere in North America, from the Arctic to California. None is found in deserts. All weasels are great drinkers and seem to require water several times a day—but just to drink. They can swim, but they don't like to, and many creatures have escaped them by taking to the water.

All members are chocolate brown on top and creamy white underneath; all are beautiful, dainty creatures—so much so that the Bavarian name for them is *Schonthierlein* (pretty little creature), the Italian *donnola* (little lady).

There is another misconception about the weasel: Its coat is supposed to magically turn overnight from brown to snow white in winter. Weasels in the north and temperate regions do have a complete moult, during which time the summer fur is shed and the white hairs slowly grow. It is not a quick replacement of one coat with another. This molt and the growth of the new white hair take about six weeks. During that period the weasels are called gray-backs; their brown hair is tinged a malt color while the new white hairs push through. The tip of the tails of the ermine and the long-tailed weasel is black; the least becomes all white.

The weasel is not all fang and ferocity. It also has an excellent brain. Sir Arthur Thompson, a British naturalist, saw two weasels (called stoats in England) encircled by many small songbirds. The birds were sitting on the ground watching the weasels jump into the air, turn flips and somersaults. The acrobats went on for a few minutes until every bird was fascinated. Then it took the two weasels less than a second to pull out of a somersault and each take a nice fat mesmerized bird.

·9·

RUFFED GROUSE

I WONDER how many of us realize that America has a little peacock, less vividly colored than the Asian bird and without the long, flamboyant, royal tail, but a bird that also unfurls its tail into a handsome colored fan, puffs out its feathers, and struts proudly. This "peacock" is never found in parks, on estates, on farms, or in zoos; it is never owned by anyone. It is the ruffed grouse, one of the most beautiful of all birds, the "king of game birds."

It also might be the strongest flyer—and the most durable.

One sunny fall day a crash in the foyer of our house, sounding like someone had hurled a brick through the window of the front door, brought me running.

A grouse sat stunned on the hooked rug, feathers hanging loose, its eyes closed. It had flown through the thick glass in the storm door, then through the glass in the permanent door, shattering glass everywhere and breaking the wooden window frames. It had undoubtedly mistaken the clear window for a fly-through space and zoomed into it.

Carefully I opened the door wide. The bird shook itself, settling its feathers back into smooth contour. Then, doing what grouse do best, it flew through the open door. The shock of breaking through two windows and their frames apparently hadn't harmed that amazing bird. It should have been dead of a broken neck.

I have since discovered that this was not an unusual accomplishment for a ruffed grouse. Not only are they adept at going through or over almost anything, but they are surviving and increasing despite obstacles like vanishing timberlines, power lines, housing developments, a multimillion-strong army of hunters, and the dire warnings of some bird experts who have been predicting this remarkable game bird's doom for years.

Several states have made grouse population surveys, with surprising results. Among these studies is one done by the New York State Conservation Department, which conducted an intensive study on two comparable grouse areas with equal cover, one open to hunting, the other a refuge.

At the time the study showed that the section open to hunting had a 39 percent larger grouse population than that of the refuge, *after* the shooting season. Predation was no factor, as predators weren't controlled in the hunting area, resulting in a much larger predator population than existed on the refuge, where they were partially under control. This seems to be conclusive evidence that New York State grouse, at least, are in superb control of the situation despite guns, foxes, hawks, owls, cold winters, and the pessimism of some bird experts.

Ornithologists call the ruffed grouse *Bonasa umbellus. Bonasa,* in Latin, means "bison," and perhaps it is a fitting name, since the wild initial flight of the grouse resembles the charge of the buffalo. Or perhaps it was allied to the drumming sound made by the ruffed grouse, which can remind an imaginative mind of the distant rumble of a herd of plains buffalo on the move. There is little doubt of the derivation of *umbellus*, which means "umbrella" and refers to the collar of the ruffed grouse, which is lifted and opened like an umbrella and is the characteristic that gave the bird its common name, ruffed grouse.

With a vast range (practically all the wooded sections of North America except the Deep South), the grouse has a variety of names, running from mountain bird and partridge, long-tailed grouse, shoulder-knot grouse, birch grouse, drumming grouse, white-flesher, tippet, and wood grouse, to pheasant, which it is often erroneously called in the South.

Separated in slightly varying color phases in various parts of its habitat, running from russet to red to gray, it is also known as the Oregon ruffed grouse, gray ruffed grouse, Nova Scotia ruffed grouse, Canada ruffed grouse, and Yukon ruffed grouse, all of which are subspecies of the same bird.

The bird of our profile, the one found in New England where it first won fame, is a handsome "peacock" of a bird with its head, back of neck, and upper body a chestnut color that is slightly spotted with white, gray, and black. It has a black-barred "crown" that rises to a crest 1½ inches high when it is alarmed, drumming, or curious. The distinct ruff at the base of its neck is formed by tufts of black, brown-tipped iridescent feathers.

It is among the largest of game birds, often reaching a length of 18 inches and weighing 30 ounces. Males and females look alike, with the female being smaller and rarely weighing more than 22 ounces. The eighteen-feather tail can be spread into a beautiful color-banded fan that is black at the bottom and red and gold as it spreads.

That tail, unfurled, is an aid in flying, becoming a rudder

that steers it skillfully through the forest at high speed. When grouse unexpectedly blast off in the woods, they have an astonishing way of putting trees between themselves and you. I have marveled at the way they dodge and twist around trees and foliage without slacking speed, never having an accident.

Birds of the World understated what happens to many an observer when the grouse flies.

> The bird crouches motionless on the ground, with which its plumage so blends as to make it invisible; and many a would-be bird watcher threading his way noiselessly through the forest, has been so startled by the sudden rushing forms springing from under his feet as to forget the object of his coming until too late . . .

The experts say that the ruffed grouse doesn't fly faster than 35 miles an hour. They claim that the bird is so skillful at flushing from behind its enemies and making that unexpected, tree-dodging flight that it just seems fast. Sometimes the grouse changes that explosive technique, taking flight from a tree, gliding as silently as an owl. It is a fact that the grouse rarely flies more than 300 yards. I have been startled by one of the birds zooming into the air under my feet and landing in the underbrush about 150 to 200 yards away. When I flushed it into the air again, the grouse spread and fanned its tail, using it as a rudder, to guide its way.

That tail is also brought into prominent display when the grouse does its famous drumming, mainly in March and April, and when it approaches the female that is lured by the sound. Naturalist Ernest Thompson Seton described it as a loud tattoo, "Beginning with the increased thump of a big drum, then it gradually changes and dies away in the rumble of a kettle drum. It may be briefly represented thus: *Thump-thump thump, thump; thump, thump-rup rup rup rup r-r-r-r-r-r-r-r-r . . .*"

Many believe the sound is produced by the male sitting on a log and beating his wings against it. But I have watched a big male sitting on the ground drumming. It was an open place in the woods that had been cleared of leaves and forest debris, an area where the grouse had perfect vision from every quarter should danger appear. But it was my land and I knew it well; I sat hidden behind two young cedars. This lusty male made his sexual salute with the sound of tom-toms. He suddenly straightened and flapped his wings, reminding me of a mallard drake trying out his flapping power while sitting on a pond. The grouse began whirring his wings in short, powerful strokes, alternately beating his sides and the air, the wing-beat finally losing its power but still fast, then the drum of sound dying in a sputter. He stopped, shook himself, then went through the performance again. By my watch he beat his drum every seven minutes for nearly a half-hour. I concluded that the sound was created by air being pushed between wings and body. Later I discovered that the experts concur, and that the beating-the-log theory is just that—theory, not fact.

But it is a fact that the grouse usually does its drumming while standing upon a log. Some naturalists believe that the selection of that log, or several logs, is so important to the bird that if all other territorial conditions are right, the grouse will not leave that vicinity for the rest of its life.

The drumming is essentially a mating call that is used by the male to attract females in the spring. But it is also sounded in the fall and summer to express sheer male vitality or to sound his defiance of other males, warning them to stay out of his territory.

Polygamous, the drumming male will mate with as many females as its strength permits, approaching them when they answer his call, his wings and tail outstretched as if for a battle. His ruff is full out as he circles, hissing. If a hen is attracted, he struts around her like a peacock, flared tail trailing the ground. Mating is preceded by the aroused male pecking the female on the head and bill.

The result of this flamboyant and noisy wooing is a nest, which is usually not much more than a dent in the ground beside a windfall tree, a rock outcrop, or a pile of brush. But I once found a female that had used an old hawk's nest to raise her brood high above most of her enemies. More usual, though, is the ground nest sparsely lined with pine needles, grass, or leaves, where she lays from eight to fourteen buff-colored eggs—one a day—some with tiny light brown spots. She does not start the incubation period of about twenty-four days until all the eggs are laid, with the result that all quarter-ounce chicks hatch from the clutch at the same time. But after a week on the eggs, she discards some, rolling them out of the nest. Biologists have examined these eggs and in every case found them infertile.

But the simple act of bringing her young into the world isn't, in fact, so simple. Both eggs and young can be eaten by snakes, skunks, raccoons, squirrels, opossums, and crows. The young and the adult are also attacked by a formidable assortment of enemies ranging from weasels and house cats gone wild to owls and hawks and those menaces to most wildlife, free-roving dogs.

After they have pecked their way out of the egg with the egg tooth, the precocious young can move out of the nest in about a half-hour, or as soon as their feathers dry. In twenty-four hours they can walk and run; in a week they can fly a few yards. In a month they are past most natural dangers such as accidental drowning or overexposure, and their flight is strong enough to get them up in a tree, out of reach of some of the predators.

That early flight strength was built by their mother steadily feeding them insects and later teaching them how to eat seeds, berries, fruits, and anything else small enough for them to catch and swallow. The female also has vocal signals—clucking, cooing, chipping, putt-putting, hissing—with which she instructs her brood when to freeze into immobility, when there is imminent danger, or even when there is insect food to harvest quickly.

Grouse appetites are flexible. They will eat about anything they can digest during the bad winter months, even dry leaves and hard twigs. Wild apples and tree buds are a favorite, especially the buds from male aspen trees. As is the case with most of our wildlife that is surviving, the ruffed grouse is still with us because it has learned when, where, how, and what to eat. Scientists have discovered that the grouse feeds on over five-hundred species of plant life and insects.

The grouse is also skillful at decoying enemies away from her brood. If she is driven from her nest or encounters danger when out with her young, she is expert at playing the helpless cripple with a broken wing, tempting the dog, cat, fox, or whatever predator is after the chicks quite a distance from her hidden offspring, then suddenly bursting off the ground in a torrent of flight just milliseconds before the enemy thinks it has her. With the chicks safe behind, she makes a wide circle, spot-checking to be sure that she isn't followed, then rejoins her family.

One ornithologist reported an unusual experience with a protective mother grouse. This bird had a single chick and tried several times with her dragging, broken-wing ruse to lure the ornithologist away from the hidden young bird.

When this proved to be ineffective, she quickly ran to the chick, seized it firmly with her bill and flew with it into the safety of the forest.

The grouse family sometimes remains together as a covey until they are mature, separating only when they are persistently hunted. I had six such family coveys of a dozen birds each on my Roxbury, Connecticut, farm for over a year.

Sometimes, however, the young, especially the cocks, will leave the family group to seek out their own home range at about sixteen weeks of age. With the dangers that lie ahead, rarely does their life span exceed twenty months, although the potential is six years. Actually, about 70 percent die within a year, killed by weather or predation. But grouse effectively fight back with a unique reproductive output that makes it possible for one pair to be responsible for 33,000 offspring in just four seasons.

Winter is the implacable foe. Even with the grouse's accommodating appetite and surprising digestive system and crop that can grind up anything from a solid nut to a piece of wood, ice and sleet sometimes accomplish what foes cannot, destroying all food supply and freezing the birds. But, as usual, they come prepared, and many survive. Nature has given the ruffed grouse snowshoes; a fine fringe of feathers grows on the bottom of its feet, and horny comblike growths along the toes distribute its weight so it can walk over snowfalls that would sink most other birds.

Snow is also often helpful rather than harmful, as the birds can fly into a newly fallen bank to burrow and rest there warmly for the night, with the snow pulled over them like a blanket. There is a danger here, but it rarely happens: During the night the snow can crust, converting what had been a refuge into a casket.

Despite the grouse's marvelous ability to combat enemies, whether they be nature or people, the bird is also plagued by something called a cycle, a marked fluctuation in their numbers occurring from seven- to ten-year periods during which they mysteriously go from many to few. Students of these game cycles, which occur in all varieties of grouse (and, in my opinion, some other birds, too, including waterfowl), believe that it is nature doing its own leveling when population gets too high for the good of the species.

Regardless of the season, if they are in sparsely inhabited terrain, grouse are in movement during most of the daylight hours. Unlike most other game birds, grouse confuse their predators by having several feeding times—morning, afternoon, and just prior to dusk. The terrain in which they range is varied. They like undulating, hilly country that is well wooded with undergrowth and broken in places with cultivated fields and meadows. Old orchards and open places are usually frequented in the early morning hours before we are up and out of doors; later, grouse go toward their refuge, the forests and the places of heavier growth, where they can hide by running ten feet or less.

They are nonmigratory and stay within a small area, usually not more than two or three miles from a favored location. So if grouse are found in a place where there is food and water, woodland and meadow, old stone walls and logging roads, it is almost certain that they will always be found there.

Another grouse rule of thumb: where there is one bird there is usually another, often several. So if one bursts off the ground with its startling sudden movement (some call this the cardiac bird), then be alert—another will be ready to tear into the air soon.

Despite its decimating cycles and host of enemies, the

ruffed grouse has been with us for a long time. Fossil study has revealed that this elegant bird, much as it is now, was flying through the forests over 25,000 years ago. No other country has such a dramatic wild bird. There are other partridges, but there is no other ruffed grouse.

Various accomplished bird experts have attempted to domesticate and breed the ruffed grouse as they have pheasants, some species of partridges, quail, ducks, and wild turkeys. But no one has succeeded yet. This is a truly wild bird that mates and reproduces only in the freedom of uninhabited forest land where the male can drum out his seductive song and proudly strut in forest shadow before an impressed female.

The ruffed grouse is a national wildlife treasure that many of us do not realize we have. A walk in a wild woodland in the fall could remedy that, if one has learned how to "see wildly." Those fortunate people who have observed grouse in their natural habitat consider it just about the ultimate in bird-watching.

· 10 ·

BALD EAGLE

THE BIRD'S WINGS touched the air like spread fingers feeling the slight wind that stirred the leaves of the trees. It was a huge bird, its dark brown body glistening in the late fall sun.

I thought it was a hawk, probably a hunting goshawk, for it was flying low, gracefully winging just above the treetops. Walking along the Mongaup River in Sullivan County in upstate New York, I was following the river, taking my time, looking for beaver signs, hoping to discover a lodge somewhere in the area.

Suddenly the large bird swooped down and perched on a branch of an oak tree. The sun glared on it like a spotlight. I stood transfixed.

The bird had a white head. Only one large, hawklike bird in America has a white head—the bald eagle. But it couldn't be. I had just read that New York State had only one nesting pair of bald eagles. Was it my good fortune to see one of those two birds? Or was it perhaps an eagle escaped from captivity, from a zoo?

As I watched, I saw that it was indeed a perfect specimen, a mature bald eagle, our national bird that has become one of our country's rarest birds as the result of industrial pollution, chemical sprays, and a misinformed public.

Knowing its plight—that it is endangered in forty-three states and threatened in six—I informed the proper authorities of this unusual sighting and received some heartening information. That eagle, they informed me, had not escaped from captivity but in all probability was wild, one of thirty-two that had been observed in the area in biweekly counts. In fact these bald eagles probably had gathered here from farther north to feed on fish which pass through hydroelectric turbines on the river.

I have seen a number of bald eagles during the past several years, but the one that I saw in 1978 gave me the most hope for the future of this embattled bird.

The noble appearance, strength, and courage of all eagles has impressed us for centuries. Roman emperors used the eagle as their personal emblem and believed that the bird would carry their souls to heaven after death. Rome's legions carried the eagle on their standards and flags as they virtually conquered the world.

The eagle was the favorite bird of the Greek god Zeus, king of heaven. It was the emblem of Russian czars, Austrian emperors, and all German rulers. But even though it is our national bird, it does not seem to have the respect of modern people. The story of what is happening to this unique bird—the largest bird of prey in North America today next to the California condor, our country's symbol of democracy and freedom—is not happy or even too encouraging. I report it here to inform readers, to correct misconceptions, and to say what must be said if the bald eagle is to have a future in America.

In late 1971 national attention was focused on the bald

eagle. Radio, television, newspapers, and news magazines reported that James O. Vogan, an aircraft pilot, appeared before a Senate subcommittee and testified that from September 1970 to April 1971, he and others had killed 770 eagles using 12-gauge shotguns, shooting the eagles from helicopters while the birds were in flight or roosting. A leading Wyoming stockman, Herman Werner (among others), paid a bounty of $25 for each bird.

Coupled with this, also near Casper, Wyoming, fifty eagle carcasses were found. The birds were killed by eating thallium sulfate that had been planted in the bodies of dead antelopes. Federal authorities tracked down the responsible stockowners and fined them. The fine was minimal, but the effect maximum and, I hope, lasting.

Those two flauntings of the law that protects our historic bird caused considerable public uproar and may have put into effect new processes of law and order and public awareness which can save the bald eagle from what looks like almost imminent destruction.

The bald eagle has been protected, although haphazardly, by law since 1940, when the bald eagle act of that year forbade the taking, selling, or killing of eagles for their parts, unless under special permit. It provided for fines of not more than $500 and imprisonment of not more than six months, or both. The preamble to that legislation was, "The bald eagle is no longer a mere bird of biological interest but a symbol of the American ideals of freedom."

On February 7, 1972, as a result of public clamor over the Wyoming killings, Congress gave that law new teeth, raising the penalty to $5,000 and imprisonment to one year.

The plight of the bald eagle was given much coverage by the media in 1972. It was spelled out that stock owners considered the eagle an enemy because they were convinced that the bird carried off young livestock of calves and lambs. In-depth scientific investigation proved that it does not, but the belief still exists.

The National Audubon Society was also doing its best to convince hunters to stop shooting all birds that they thought were hawks. As this is written, many hunters take any opportunity they can to shoot hawks (even though they are protected in most states), believing that these birds are the principal killers of game birds and rabbits. It is untrue, but most hunters remain unconvinced.

The major problem of the bald eagle is that the gleaming white head and tail feathers, the marks by which anyone can identify the bird, do not appear until the eagle is more than four years old. Immature bald eagles are completely brown and look exactly like what many believe to be harmful hawks.

But an even larger problem concerning the eagle's survival is the increase of human population in its primary nesting areas, causing disturbance of nesting birds, loss of nest trees, and our still-growing giant killer, pollution.

Pesticides in the eagles' food cause them to produce infertile eggs. Now mercury has been added to the other slayers of the bird. If they can get it, 90 percent of the bald eagle's diet is fish. Industry is dumping waste mercury into our waters, infecting all marine life. Recently dead eagles examined at the Department of Interior's Patuxent Wildlife Research Center in Maryland were found to contain lethal amounts of mercury. Canada admits its industry pours 250,000 pounds of mercury yearly into the Great Lakes alone. The United States doubles that. And mercury is a persistent killer, since it remains in our lake and stream bottoms for from fifty to one hundred years.

The result of all this? Once abundant in all states, numbering in the hundreds of thousands, there are now less than two thousand nesting eagles left in the United States (exclusive of Alaska). Eagles have always thrived along the coast of Alaska and near its inland streams, where fish are plentiful. But now the bird is threatened even there. Canning factories have nearly wiped out the salmon, a fact belatedly realized just after World War II. Yet the steady decline of salmon was blamed on the bald eagle. Records show that Alaska paid two-dollar bounties for a pair of the clawed feet of 128,273 eagles. Many thousands of the birds were also indiscriminately killed without bounties being paid. It wasn't until 1952 that the slaughter was stopped by law.

The real tragedy of the bald eagle is misunderstanding. Biologists and naturalists have known for a long time that the bird's activities are helpful to us. Being sea eagles, they feed on fish, but mostly fish we do not eat. Herring and salmon that they do kill have proved to be mainly those inedible ones that have spawned and are ready to die. Eagles also have been accused of killing game birds and wildfowl. They can't kill game birds because, with their more than seven-foot wing span, they cannot maneuver in the woodland where these birds seek refuge. Occasionally they may take an unwary duck or coot, but the eagle is no match for a healthy, swift, dodging duck. One scientist saw a bald eagle swoop in and drive off a flock of gulls fighting over a large dead fish. The eagle gracefully swept in, picked up the fish and landed on an ice floe, then ate the catch. What was interesting to the observer was that the eagle flew low over several flocks of rafted ducks without trying to harm them and, more important, without even alarming those ever-wary wildfowl.

Let's look at the life-style of our national bird to learn how it became famous and perhaps clear up a few more misunderstandings.

After our thirteen colonies declared independence from Great Britain at the Second Continental Congress in Philadelphia on July 4, 1776, it was decided that the new government needed an official seal. Benjamin Franklin, John Adams, and Thomas Jefferson were appointed as a committee to prepare one. First they offered a head of George Washington and the words *E pluribus unum* ("one out of many"), which was rejected.

Nothing was decided until 1782, when Charles Thomson, secretary of Congress, took a heraldic eagle design by William Barton and substituted our own bald eagle holding the slings of war—thirteen arrows—in its left claw, and the olive branch of peace in its right. In its bill was a scroll that was inscribed *E pluribus unum,* as suggested by the first committee. The other significant items on the seal were the thirteen stars and a three-color escutcheon covering the breast of the rising eagle, with its wings extended. The three colors of our flag were used: red for hardiness and valor, white for purity and innocence, blue for perseverance and justice.

Even before this seal was designed, the bald eagle flew in America on the New York State crest. At present the bald eagle is prominent in the armorial designs of Alabama, Arkansas, Illinois, Mississippi, New Mexico, Oregon, Pennsylvania, and Utah.

Benjamin Franklin wasn't happy about the Great Seal being designed around the bald eagle, saying that the bird was a bully, a robber that stole from other birds and an eater of carrion. He suggested that the wild turkey would be a better symbol. In his comprehensive study of the natural and civil history of the bird, *The American Eagle,* Charles Hobart Herrick put Franklin straight.

> The bald eagle is a true native son, being found only on the continent of North America, which he has never been known to leave of his own volition. He nests high, as near to the sun as he can get. He is a model parent and probably spends more time—upwards of six months—in rearing his family and giving his progeny a fair start in life than any other bird known to this continent, if not to the world. He does not live entirely or mainly by "robbing and sharping" as Benjamin Franklin seemed to believe, but is an expert fisherman in his own right, and will not rob the osprey unless this bird is heedless in giving him the chance, or offers him a challenge. The eagle is no coward. He is never driven from his neighborhood by any living being except a man armed with a gun . . .

This is mainly accurate. But the bald eagle is afraid of the great horned owl and has been driven from its territory and even killed by this large, fierce, night-raiding bird. And the eagle does occasionally bluff that skilled fisher, the osprey, into dropping its catch, then swoops beneath and retrieves the fish in the air.

The bald eagle is indeed a fit national symbol. Imposing in appearance, with its lordly white head and brown feathers and markings, and majestic in flight, it was once seen across the length and breadth of the land and was truly a "national" species. Its range was from northwestern Alaska eastward through northern Canada to Newfoundland and southward across the continent to lower California, Arizona, New Mexico, and the Gulf states.

There was a somewhat smaller southern bird that nested primarily in the estuary areas of the Atlantic and Gulf coasts, from New Jersey to Texas, and the lower Mississippi Valley southward from eastern Arkansas and western Tennessee. Some of these birds wandered northward in summer after the nesting season to the northern United States and southeastern Canada.

The northern bald eagle was most plentiful in Alaska and Canada, although it sometimes also inhabited the northern states, particularly around the Great Lakes. These northern birds also migrated in the winter, flying to areas along the Mississippi River and in Oklahoma.

Today its only safe flight is on the President's flag. It also flies free on our highest government award, the Congressional Medal of Honor, is engraved on the Navy Medal of Honor, the Distinguished Service Medal, and the Distinguished Service Cross, and perches on the lapel pin of those who are honorably discharged from military service. The bald eagle emblem even flew to the moon on the mission patches of Apollo astronauts.

But these honors proved no armor at all for our beleaguered national bird. Its enemies either did not know of the eagle's prestigious place in the American way of life or just did not care. They claimed—and many still do—that the bald eagle's appetite makes it a menace.

Untrue. The eagle's diet is catholic and not competitive with ours. If the bird isn't successful in obtaining its favorite food—fish (usually so-called trash fish)—it will eat anything it can catch, from a rat to a rattlesnake. And if it can't find live "natural" prey, it will make a meal of carrion, such as deer that have been run to death by free-roving dogs.

Stories of eagles killing and carrying off fawns and young livestock have never been proved, despite much investigation by objective scientists. Biologists conclusively demonstrated that a bald eagle cannot lift more than its own weight off the ground and can barely manage that. As a mature eagle rarely weighs more than eleven pounds and usually averages three pounds less, it would be physically impossible for this bird to engage in the activities of which it is accused.

Also we can add to the bald eagle's assets of dignity, pride, courage and fierce independence one more asset: perfect parenthood. The eagles mate for life. The female, who is not only larger and fiercer but also the wooer, incites the male into the twice-daily love play. In *Autumn of the Eagle*, George Laycock, quoting another team of writers, writes of that courtship of the pair of eagles that have become inseparable at the beginning of the mating season.

> They soar out of the afternoon thermals and rise together so far above the earth that the human eye sees only two distant specks among the clouds. Then, with the rushing winds slipping over their wings, and their bodies under perfect control, the bald eagles sense their power and taste the full excitement of spring. Suddenly one of them rolls over and, flying upside down far above the earth, extends its curved yellow talons in an invitation to its mate flying directly above it. With feet extended, the birds touch, their talons intertwine and lock together, and the bird on top rolls over, trading places with its mate. They then cartwheel on set wings over and over toward the earth. Up and over they pinwheel down the spring sky until it seems that they will crash to earth. But then they break apart, and, on heavy wings, lift themselves to the security of the eagle's sky again.

After that spectacular wooing a nest is constructed by both

birds, often a hundred feet up in a large pine near water. Sometimes it is built in four days of strenuous work. The pair will return to the same nest for many years, adding to it, renovating it. There is a record of one pair of bald eagles nesting in the same eyrie at Vermilion, Ohio, for thirty-six years. The nest at the end of that time weighed two tons, was ten feet in diameter, and strong enough to hold a person.

But it is designed to hold two or three (sometimes four, but rarely) bluish-white eggs, which hatch in about a month. The eggs are small for such a large bird, measuring 2½ inches long and 2 inches across. Both parents incubate the eggs. Hatching takes about eighteen hours for each egg. The birds are born at different times, so that often the last one hatched doesn't compete successfully for food and sometimes dies or is killed by the stronger nest-mates. Because of that and other factors such as infertile eggs, raising the young successfully is reduced to 48 percent.

The young, with smoky-gray down, are in the nest for twelve weeks. When they are a fully feathered dark brown at a little over three months, they are taught to fly by the parents, who hover just above the nest, offering tempting food. Once the eaglets are lured aloft, the parents spiral, forcing the young to follow, and giving them a concentrated aerial exercise before returning to the nest to feed them. Eaglets are fed eight times a day.

From that first testing of wings, daily test-and-instructional flights are made. Hesitant young are pushed out of the nest. Protecting the young with vigor, parents keep the eaglets with them for six months, teaching them to hunt and fish, to dodge and twist in the sky, and to build a nest.

At the end of their tutelage, the young eagles can spot prey on the ground from a distance of three miles, can dive at speeds up to sixty miles an hour, and rarely miss a target. With a hooked, gold-yellow beak for tearing flesh and one-and-a-half-inch talons for capturing and killing their prey, bald eagles are well endowed for their role as fish or sea eagles, as they are often called.

Our bald eagle, one of two eagles in North America (the other is the golden), belongs to an impressive family, which ranges in size from the little eagle of Australia and New Guinea to the 20-pound harpy of the rain forests of Central and South America. Most unusual in the clan is the large monkey-eating eagle of the Philippines.

The bald eagle is classed as a diurnal, or daytime, bird of prey, as are the 280 species of other hawklike birds, which include all hawks and more than 50 species of eagles, vultures, falcons, harriers, buzzards, and kites. There are just 8 sea eagles, which live near large bodies of water and whose diets are 90 percent fish. The largest is the 19-pound Steller's eagle of the Aleutian Islands.

Actually, the eagle is lucky to leave the nest at all. The average number of young produced under normal conditions is only 1.4 per active nest. Of those that do make it to maturity, the female will be larger than the male, perhaps measuring 42 inches and weighing an average of 10 pounds, while the male will be about 35 inches and weigh an average of 8 pounds. The female's wingspread can measure 8 feet against the male's 6½.

Scientists call our eagle *Haliaeetus leucocephalus*, which translates into "white-headed sea eagle." The further translation into bald eagle probably came from a Greek word, its original meaning being white, not bare, and a Latin word meaning head.

Our national bird has other names—American eagle,

Alaska eagle, black eagle, bird-o-freedom, white-headed eagle, sea eagle. The immature bald eagle is sometimes called a gray eagle.

If it can escape guns, learn to detect chemically tainted fish, and find a place to nest in an area that we don't covet, the bald eagle can live for forty years. But few seem to be fortunate enough even to last four years, when they acquire that gleaming white head that marks them so prominently.

Recently 10,788 acres of the Tongass National Forest in Alaska were designated as a sanctuary for bald eagles. As southeastern Alaska has the greatest concentration of eagles in North America, and the primary purpose of this sanctuary is for studies of our vanishing bird, the move is heartening. The U.S. Forest Service is also setting up protective sanctuaries and trying to work out a management program that will include inventories of nests and improvement and protection of nesting areas. Also, the powerful National Wildlife Federation is working to establish a National Bald Eagle Refuge.

There is no doubt that people and eagles can live together. A dentist on Marco Island in Florida has a pair of nesting bald eagles in his backyard. All involved seemed contented.

The eagles in Alaska are now laying eggs that contain only two parts per million of DDT, a persistent killer of the birds because it makes their eggs infertile. This is heartening; as of this writing the Alaska eagles are successfully raising their young. In the extreme southern part of Florida, mostly in the Everglades, a healthy population of eagles continues to reproduce, and eagles are still nesting in the 1½-million-acre Chippewa National Forest in Minnesota.

But the three "killers" of our national bird are still loose in the land: pesticides, guns, and vanishing breeding areas.

We must have faith in the future and in the determination of good people. Whooping cranes, which were once numerous and were killed in such numbers that only a couple of pairs of birds remained, have since made a successful comeback. Those birds are now reproducing. They are protected; they are guarded; the public knows their plight and has responded.

If we can save the whooping crane, we can save the bald eagle. If we do not, our new national symbol should be the dollar sign.

· *11* ·

LOON

I DIDN'T SEE my first loon, I heard it. That call of the wild has intrigued people for centuries. Many Scandinavians believe that the sound of a loon can predict or even bring bad weather, and that when this mysterious bird cries in the night, it is the voice of water spirits telling of a human drowning to come. Medicine men and women in the wilder regions of Russian Siberia dangled loonlike carvings around their necks, which were supposed to help them make the magic of sending their spirits to the world beneath the water. Many nationalities believe that a person who kills a loon will soon face disaster.

I knew none of this when I heard my first loon cry in the lonely night one summer years ago, when I had fled the city. The scene of that first meeting with the sound that introduced me to one of the most unique of wild birds was in itself unique: An Adirondack Mountain lake, night coming down, dropping long shadows on the water. The campfire was dead, the smoke still lingering in my nostrils. White pine needles dropped from the trees onto the canvas of my tent and lulled me to sleep as gently as soft rain.

The high, weird call brought me upright in my sleeping bag.

It was a pure, primeval sound that makes even the wild bark of the migrating Canada goose seem tame in comparison. Again the cry tore through the night, ripping apart the silence that had settled on the lake. It was answered from the far shore of the lake. Then it was quiet, and I lay in my sleeping bag waiting for it again, thinking that this really is what I came here for—the wild sound that in an instant transported me from my mundane life into that old natural one that is eternally mysterious.

Those strange, high notes hurling across the dark water did more to set me in time and place away from city streets than the long hike up through the quiet forests, more even than the first sight of a hawk making its incredibly graceful swoops. That loon sound ran in my blood even as sleep stole over me. That was the door through which I left civilization and entered a world that still strives to exist naturally and untouched.

I am one of the fortunate ones who has heard the chorus of Alaskan wolves; their song is an instant spine-stiffener. But for me, loons, with their many different notes—beginning low, rising high, then suddenly dropping—are an even more poetic symbol of what wildness is all about. Those calls have been variously described as defiant, sinister, mournful, mirthful, uncanny, demoniacal. And if you've never heard loons on an isolated lake just before a storm comes up, then you have missed one of the outdoor world's most stirring sounds. They start with lonely, plaintive low cries, then rise almost to a cry for help, then level out into a witching call that seems to be orchestrating into control all the wild elements out there gathering into a storm. Other loons answer, and then the storm breaks—lightning and thunder, driving rain, and strong wind.

Naturalist Robert Hegner claims that the cry of a loon leaves a lasting impression, as though in hearing it you had finally reached the primeval wilderness. Canadian wildlife expert P. A. Traverner says that many people get more enjoyment from loons than from fishing, and that there are hundreds of outdoor recreation areas where loons provide the finishing touch of native wildness. According to writer Hugh Fosburgh, a loon is something more than a bird, and is a creature that would have fascinated Edgar Allen Poe. "They are the evocative essence of the northern wilds," he wrote in *A Clearing in the Wilderness.*

Pinning that evocative essence down with hard facts places the loon in the order *Pygopodes*, suborder *Cepphi*, and the family *Gaviidae*, of genus and species *Gavia immer.* They have changed not at all, from the evidence of the fossil remains from the lower Miocene era found in France and from the Pliocene era found in England, which means that loons are old, dating back millions of years, and that they were originally well made for survival. Four living species are recognized. The one most of us know—if we know the loon at all—is the common loon, also known as the big loon, great northern diver, and hell-diver. Close relatives are the Arctic and the black-throated and red-throated loons.

The common loon is a poor name for a most uncommon creature. The loon is a large, striking bird with a black-and-white checkered plumage and white belly. Its head and neck are black with a purplish iridescence and vertical white bars that form a necklace. In size it resembles a goose with a short neck. Adults run from 28 to 36 inches in length and sometimes weigh as much as 12 pounds. Males and females look alike, although the male is somewhat larger.

The Eskimos have a fanciful legend explaining how the loon

acquired its handsome necklace: A blind boy wandered to the water's edge and a loon guided him onto its back. It dove and swam underwater three times, the boy clinging to the bird's back. When they rose to the surface for the last time, the boy was no longer blind. As a reward for restoring his sight, the Eskimo boy gave the loon its jeweled necklace.

Loons make their summer residence from New England and the northern edges of the United States to Labrador and Newfoundland, and they migrate through the interior states and the coasts to the Gulf of Mexico. Thus they are familiar to many who frequent the waterways throughout North America. Breeding, however, is usually restricted to the unspoiled areas of the northern lakes.

Bloody battles ensue during that breeding season, with competing males using their long sword-beaks as skillfully as fencers, but more seriously. Fierce fights have been recorded in which a victorious loon suitor has actually severed the top of another's head before swimming off to claim its prize, the waiting female.

The versatile cry is also used in mating. Loons have been heard crooning to one another, sending plaintive love calls across the water. I watched the remarkable courtship water-dance of one pair. Unlike most bird species, in which the male alone does the display courting, both loons participate. Pulling themselves erect as if they are held out of the water by an invisible string, they stretch almost to a three-foot height and gyrate across the water, seeming to dance on its surface. They also move around one another in a "circle dance," raising and lowering their beaks in what is called bill-dipping.

The result of these mating rituals is a serious alliance that lasts for life. It also produces a pair of eggs shortly after the wooing is over. Uncommon in about every other aspect, the common loon is very common about its nest, which can be a crude affair of moss, grass or cattails, and mud in a mere depression in the ground, atop an abandoned muskrat lodge, or on large pond-lily roots. But it is always located near enough to water so that the nesting loon can slide off into it. The nest also is situated so that the loon has unrestricted vision and can spot danger instantly.

Although the eggs are vulnerable to raccoons, minks, and muskrats, they are neatly color-camouflaged. Usually two dark olive-gray, brown-stained, black-spotted eggs, which easily blend into the background, are laid. The eggs hatch in a month after being incubated by both parents. The downy, gray young leave the nest in twenty hours and swim immediately. The loon is an excellent parent, carrying the pair of babies on her back for about a week before giving them advanced swimming lessons. The offspring grow slowly and remain with the parents throughout the summer, picking up training that will serve them after they leave the home lake or pond.

The young loons can dive and catch and swallow fish instinctively soon after they leave the nest, but watching their parents fish and hunt perfects their technique. They can actually swallow small fish underwater, helped by their unusual tongues that have spines leading backward into the throat to aid in swallowing the fish.

I have watched two loons and their young run races on a woodland lake as feathers of early morning mist drifted, giving an unreal, dreamlike effect to the incident. Using those wide webbed feet for propulsion and half-extended wings as support levers, the four loons pushed off from the shore together, racing down the lake, then wheeled around and scurried back to the starting point. Most naturalists believe that this is more than fun and games and that the racing exercise strengthens the young birds for the long flights ahead.

The method of flying is strong and distinctive once it gets underway. The bird looks humpbacked, with its head lower than its body and its feet straight out behind. Some naturalists claim that the adult loon can't get off the water at all without an assisting wind. Its compact body is heavy, and its wings, although they are extremely strong, are short. When you watch a loon trying to get off the water into the air, the effort looks so extreme that you automatically start rooting for it to make it.

Flapping the short wings, it runs along the surface of the water like a clumsy amphibious aircraft, its feet splashing water, and finally lifts laboriously off after using a large expanse of water as a runway. It doesn't gain altitude fast, but gains it by circling, finally getting aloft. But with proper altitude the loon is fast and strong, flying straight as an arrow. Its wing beats are rapid and distinctive. There is no question of identification with a loon.

Coming back down on water where it is an undisputed master, the loon plummets in a spiral from high up, hitting the water like a thrown object, landing with a splash, and skidding along the surface until water resistance stops impetus. Here it is in perfect control; great northern diver is a much more appropriate name than common loon. The loon may be the greatest diver of all the water birds. All its physical aspects attest to this prowess. Its torpedo-shaped body has legs placed way back toward the tail, which send the webbed feet forward like propellers through water. It is among the very few birds with solid, heavy bones, which is an asset in its submarine quest for fish. Its bill, which is long and sharp both at the tip and the edges, is an efficient weapon that permits the loon to spear a fish and hold it after it is caught. Its plumage is so dense and compact that water cannot penetrate it, and a thick layer of fat under the skin retains body heat.

A dive of fifty feet is routine for loons. They have been caught in fishing nets well over a hundred feet down. They also are such fast swimmers that no fish can escape them. Once, through binoculars, I watched a loon underwater in a clear mountain lake. It used its legs in a pumping motion and stroked with its wings, which made the bird look like it was actually flying through the water—which it nearly was, it was moving so fast. Although the loon holds its breath underwater and usually surfaces in about a minute, loons have remained submerged for five minutes while traveling nearly five hundred yards. Often they just stick their heads underwater to look for fish, with a vision that some scientists claim is eight times as acute as ours.

As their food is mainly fish (with aquatic insects, crayfish and other crustaceans, freshwater mussels, frogs, and other small amphibians filling in), some sportspeople claim that loons are so good at fishing that they take too many game fish. Biologists have disproved this by showing that the loon's gullet is too small to handle large fish. Overfishing frequently is blamed on everything and everyone except the person with the fishing rod and net. Loons do not work the swiftly moving streams where trout are or the marshy shallows that are the habitat of some game fish. Birds that have been examined were found to have been dining mainly on fish such as chubs, shiners, and dace.

Although it is perfectly designed for its water habitat, the loon is vulnerable on land. Because their legs are placed so far back, loons walk awkwardly and must rest in a peculiar, upright position. In motion on land, they look crippled. Loons have been found in open meadows where they had set down in a dense fog, mistaking it for water. Helpless, they were easily caught and placed in their watery element. Yet I once saw a wounded loon outrun a man for fifty yards. Desperation probably gave it the speed. Other loons, trapped on land, have turned and fought aggressively.

Their aggressiveness isn't keyed only to defense or mating. One observer spotted a third loon that had landed on a pond where a pair had been in residence. It was near the shore and acting strangely. The man got into a rowboat and went to take a look. The loon didn't flee as loons always do. It seemed tame for such a usually wary bird. Thinking that it might be the offspring of the pair in the middle of the pond, the man gently herded the loon over toward the other two. Then he watched in horror what his action had precipitated.

Screaming madly, both of the other loons charged, one viciously attacking the third loon, hammering at its head with long, sharp bill, knocking it sprawling on the water. Then the attacker pulled the weakened loon's head under the water, obviously trying to drown it. When the man rowed over to break up the fight, the aggressive loon pulled the other completely under water, forcing it near the bottom. All during the battle, the third loon sat on the water uttering a high, devilish cry as if urging its mate on.

Finally, as the man watched, both fighting loons surfaced, one seemingly dead. The victor made a loud cry, joined its mate, and swam to the other side of the pond. Picking up the inert loon, the man placed it on the bottom of the boat. As he rowed toward shore, he noticed that the bird blinked its eyes. But by the time he got to shore, the loon was dead. A field naturalist, the man took it to a biologist friend to have it autopsied. The result shocked him. The dead, severely mauled loon was a female. The time of the fight wasn't the breeding season. So he had questions: Was the bird that attacked also a female? Would a male loon attack and kill a strange female? Why? The man now realized that when he first saw the third loon acting so lackadaisically, it obviously had been recovering from an earlier attack by the loon that had finally killed it.

Loons don't restrict this deadly dueling to their own kind. I once spent a month on a little lake tucked away in the fastness of New Brunswick, Canada. A pair of handsome loons were in residence. They kept me amused and enthralled with their antics and their lyrics, flying around the lake in the hours before dawn, trying, it seemed to me, to screech the sun up over the mountains. I also learned about the reason for the phrase "crazy as a loon." In the morning, after they had their daily and loudly vocal exercise flying around the lake, they would set down on the water and go about getting breakfast, diving deep and nearly always surfacing with a small fish. After they ate the fish they would burst into loud "laughter"—senseless, purposeless laughter, according to some naturalists. Crazy laughter. But after watching and listening for about thirty days, I came to the conclusion that it wasn't crazy after all. It was sheer lusty laughter in the joy of living and being able to own a lake and fish whenever they wanted.

And own that lake they did. One day a small flock of mallards dropped out of the sky onto the lake, showing their usual beautiful form and grace. I like to watch mallards, for they are such contained, elegant birds. Not these, however. Shortly after the ducks arrived, my pair of loons seemed to disappear, and the mallards one by one began to act as if they were in pain. First a big drake screeched and lifted up off the water. Then his mate, a large Suzy, squalled and flapped. Within minutes each of the dozen ducks seemed to have been struck by some hidden force in the water that was agitating them. They had been. My loons were doing their deep and long-distance diving, attacking the mallards underwater by spearing them with their long, sharp beaks. Less than half an hour after they arrived, the mallards rose, quacking, and took off, leaving the lake to the loons, who were smugly sailing around its surface.

Loons leave northern lakes in alternating flocks (which is about the only time they associate, except when ice forces them to share open water), some going south in September, others in October, a few remaining until hard ice forms. They move south along larger interior rivers, and they also use the Atlantic flyway along the East Coast. Very little, however, is known about loon movement or total population, and their migration routes are not completely charted.

In late fall 1963 scientists therefore were surprised when seven thousand loons were found dead (thought killed by Type-E botulism) along the Lake Michigan shore during one fall migration from Canada to the Gulf of Mexico. Several couldn't believe the high death toll, saying they didn't think there were that many loons anywhere in North America. After the discovery of the dead loons, it was thought that Lake Michigan might be a natural flyway where loons met in migration from the north, northeast, and northwest, or that drought in the prairie provinces made it necessary for the loons to find a new migration route. There had never been a count of loons in the country north of Lake Michigan, nor were large migrating flocks ever seen. So the death of the seven thousand birds remains a mystery ripe for speculation, since no one knows where or how or why that many loons got together, or exactly how they died.

Loons are not a game species and are protected by the Migratory Birds Convention Act in Canada and the United States. But what really protects this picturesque species is that it seeks out the lonely places and the undisturbed lakes—often in the uninhabited far north—so its survival seems assured.

· 12 ·

GREAT HORNED OWL

THE SKULLS LAY in a pile, bleached white as pearls, the sun striking splinters of light off the staring, empty eye sockets. The forest wind stirred the leaves of the tall oak. The executioner had been at his grisly work again. A man poked a stick into the skulls, counting them.

One hundred cottontail-rabbit skulls had dropped from the high nest of the great horned owl, a bird called many things by many people, including the tiger of the skies, which it is, and misnomers such as the executioner and the devil-a-wing. The man making the death tally was a natural scientist at his task of proving that cottontail rabbits make up to 41 percent of the owl's diet.

Even scientists who have been trained for the long view can be short-sighted. In the report which was published, the owl was described as the main predator of rabbits. This is not a fact. The main predator of the rabbit is the human being. Each year in the United States, hunters bag about 20 million cottontails. It is a good thing that they do, and that owls assist. It has been expertly estimated that if all the offspring from a single pair of cottontails survived for five years, there would be more than 350,000 for each original pair. It also has been reliably computed that in Missouri alone the rabbit's potential number is 6 billion in three years.

Yes, the great horned owl eats rabbits, but another scientist, after involved study, said, "Like all predators, the Great Horned Owl takes what is most readily available. Other things being equal, cottontail rabbits would top its list. Usually other things aren't equal, however, for it is an exceptional situation when mice don't make up the major part of the owl's diet."

The rabbit report that I had recently reviewed, by the naturalist who seemed to be out to prove that the great horned owl was rapidly decimating the cottontail population, turned my thoughts back to another time when I had first discovered what a unique bird that feathery-horned owl is.

Noticing unusual droppings under a towering white pine several times, I reported it to my naturalist father, who, as usual, asked for a precise description. The nearest that I could come was that it looked like a tobacco cud that had dried in the sun. He suggested that I get several of the droppings.

Eager and mystified, as only a ten-year-old can be, I made the two-mile walk to the big pine and was back in short order with four of the tobaccolike objects in my pocket. Close up, they looked like unpressed brown fur.

Placing the 4-inch-long, 1½-inch-diameter wads on a piece of paper, my father took out his pocketknife and sliced one open. Out popped a tiny skull, white as ivory. Careful incising of each wad brought the same result: bleached skulls.

As my father identified each skull, this became more exciting for me than any suspense story. The first skull was probably a bluejay because it was fairly large, with a long beak and no teeth. One was a meadow mouse, the largest of our mice, its skull heavy and wide, with two long, chisel-shaped front teeth. Next was a mole, identified by its pointed skull and slender, insectivorous teeth. The smallest skull was a shrew, also pointed, with sharp, brown-tipped teeth.

My father, using the tip of his pocketknife to mark his facts, then explained what that mysterious brown wad was: a pellet from a great horned owl. Owls catch their prey with talons, crush it with their beaks, tear it into pieces, then swallow it—feathers, hair, bones and all. About seven hours after feeding, the indigestible parts form into a compact mass in the stomach that is disgorged in a process that takes about five minutes.

The most important lesson I learned that day, however, was that the evidence in the pellets was clear: The great horned owl is not the villain it has been painted as being. Although one pellet did have the skull of a bird, the rest contained remains of rodents, our enemies.

Since that long-ago day I have examined many pellets of *Bubo virginianus virginianus*—the great horned owl—which is one of the largest, fiercest, and probably least understood of our birds of prey. Pellet by pellet, observed and researched fact by fact, I have built a personal profile of this master of all winged hunters. A pair of resident owls live on our property, and I had a pair on the acreage we occupied before we moved here. Not only do I know them by those distinctive and revealing pellets, but they themselves tell me who they are and where they are every week, summer and winter.

Coming out of the dark night, deep and resonant as beats on a bass drum, the uninflected hoots—*hoo, hoohoo, hoo, hoo*—usually are five in number. I have learned to distinguish male from female, as the female's voice is a couple of notes higher, shorter, and not so rich or drawn out. To me, the call of the great horned owl, when it is heard from far away, sounds exactly like the cooing of the mourning dove. When it is nearby, the sound is not always a hoot, however. Barks, grunts, and startlingly loud idiot laughter sometimes pour from our woods. Thoreau also was impressed, writing that the great horned owl's hoot rang far and wide, occupying the spaces rightfully with its grand, primeval, aboriginal sound.

This owl reacts to imitations of its calls. I'm not much good at hooting like a great horned owl, but those who are have discovered that the calls are used to mark out zealously guarded territorial lines. Grossman and Hamlet report in *Birds of Prey of the World* that ornithologist Loye Miller took his students on a trip to California's San Bernardino Mountains. Hearing one great horned owl hooting, which he named Bubo, Miller mimicked the owl and was answered after each hoot and followed closely as he walked through the bird's territory.

Crossing an invisible line marking the beginning of another owl's region, Miller's mimicry was answered by another, higher hooting. This great horned owl he called Tecolote. This owl also followed Miller, answering him. As the bird expert turned around and walked back into Bubo's territory, he used his full imitative repertoire to try to coax one owl into the other's area. But neither bird would fly into the other's territory. Miller told his rapt students, "This no man's land between established territories held by Great Horned Owls allows them to live in comparative peace, instinctively, and without recourse to treaties."

Those booming hoots stake out a large territory that runs,

experts believe, as much as six square miles per bird or pair of birds. The calls also are dramatically used in the mating ceremony. Beginning in December the hooting changes in tempo to an oft-repeated *whoo, hoo, hoo, hoo, whoo, whoo*. There is a rising excitement in the male's call, which is sometimes answered by the shriller notes of the female. When he has attracted her, he fluffs his feathers, bows his head, spreads his wings, and even does a little shuffling dance. Capping the repetitive courtship ceremony with aerial acrobatics, he ground-loops, spurting from tree to tree, going high, then tumbling in a deadfall from the sky. If she is sufficiently impressed, the female, which is larger, responds by bowing and fluffing her feathers and hooting a little. They snap their beaks at one another in a sound not unlike castanets. Finally they get together on a broad limb of a high tree, enthusiastically rub heads and bills together, then stand back facing one another and hoot out a duet. They mate for life.

Rarely do great horned owls build nests. As lords of all they survey in the forests, they simply confiscate a nest from a hawk (particularly the red-tailed) or a crow. Often, though, working together, they will clean the debris out of the borrowed nest, and the female will pluck out its breast feathers and line it. During late January or February she lays two or three large, dull white eggs at intervals of several days.

As this is the height of the cold season, the female must closely hover over the eggs to keep them from freezing. Even though bird and nest may be covered with snow and ice, she remains on the eggs, her body heat keeping them and the center of the nest warm. The male supplies her with food and roosts nearby to defend her if the need arises. The only real enemy the owls have is human, and even this most dangerous predator has frequently been foiled. There are documented cases of horned owls savagely attacking and wounding people who climbed a nest-tree.

In about thirty-five days, chicks—downy-white, weak, unable to hold their heads up, and blind are hatched. At two weeks the down is replaced with a mottled buff. The chicks remain in the nest for a month and a half, with both parents diligently feeding them. One observer saw the male and female make fourteen trips to a nest in less than an hour, bringing field mice, rabbits, rats, and muskrats.

The young can't fly until about the twelfth week, when the contour feathers are grown. Although the rearing of young horned owls takes about five months—twice as long as it does for the big hawks—by nesting in the dead of winter the horned owl has the important advantage of time over every other winged predator. In late March and early April migrating birds return, and everything from the later-nesting songbirds to hawks and other owls are a food supply for the horned owl's young.

Also in mid-May when the immature horned owls are ready to try the complicated business of survival on their own, the young of most other wildlife are there for the taking. And owl parents have had the time to patiently teach their offspring how to hunt, testing them on everything from snakes and mice to frogs. Although they will not become masters of the art for some time, by winter the young are proficient enough to feed themselves.

As I said, that feeding causes controversy and misunderstanding. Everything smaller than a dog that moves is fair game, including those clever predators, young foxes, weasels, and minks. Skunks are a favorite meal for horned owls, their burning, blinding spray apparently being ineffectual against them. The porcupine, however, has been known to drive its quills into an adult horned owl and kill it.

With this completely catholic taste, the horned owl takes what is available—rabbits, game birds, even poultry if it is unfenced and uncared for. Scientific tests, however, have proven that the bulk of the horned owl's diet remains mice and rats. This is significant: Consider that in one year a single pair of meadow mice is said to be responsible for the birth of one million destructive mice.

One biologist, attempting to prove that the horned owl really is our friend and that the bird takes mainly the most available prey—rodents, selected a horned owl's nest at random and tallied his findings. The nest contained the remains of 1 mouse, 1 muskrat, 2 eels, 4 bullheads, 1 woodcock, 2 grouse, 1 rabbit, and 11 rats. Another in-depth study in one section conclusively proved that the owls took nearly 25,000 rodents for each square mile.

Strengthening this thesis further, Ronald Austing and John Holt, authors of *The World of the Great Horned Owl*, reported that when food is abundant, the owl dines on just the preferred portions of its prey, thereby pointing up what that favorite food obviously was. E. O. Niles, they said, told of a nest of owls housing two young birds and several dead rats. On the ground under the nest were the almost untouched carcasses of 113 newly killed rats with just the skulls opened and the brains removed.

Our helpful horned owl has been around a long time—about 36 million years. Fossil remains prove that it has changed little, if at all, since the time of the Cenozoic era. With a length varying from 18 to 25 inches, a wingspread that can reach 60 inches, it is distinctive. It has ear tufts or horns and a conspicuous white collar, and comes from a large and complicated family numbering 500 types. Ranging from the great horned owl to the sparrow-sized elf owl of Arizona, relatives live in nearly every part of the world; there are forty-five species of owls in North America. Leading in importance, the great horned owl occupies southern Ontario and Quebec, western New Brunswick and Nova Scotia south to the Gulf Coast, Florida, and west to Wisconsin, through Minnesota, South Dakota, Kansas, Oklahoma, and western Texas. Other species of horned owls occupy a much wider range, including the Arctic.

The color phases of the horned owl number approximately ten, and vary depending on where the owl lives. Ornithologists break them into subspecies ranging from the subject of our profile with its dark plumage, dusky and tawny markings, and rusty face, to the paler western horned owl, through the tenth species, the dark- and the light-feathered dwarf horned owl, which is found only in the southern section of lower California. Regardless of species, all horned owls are endowed with physical assets that make them the masters of winged hunters. Their eardrums are the largest of any bird. The face is surrounded by a stiff curve of feathers that bounces sound waves into its ears. So keen is its hearing that it can locate prey in darkest night entirely by sound.

Its eyes are the most perfect organs of vision of any creature. One hundred times as acute as those of humans, this owl's eyes can instantly be focused to see near or far (in one test an owl captured its prey in light so dim that it was equal only to that of an ordinary candle burning 2582 feet away). Set in front of the skull, the eyes have what is called binocular vision and cannot be rolled from side to side. But the owl compensates for this by rotating its head 180 degrees in either direction and can look directly back over its shoulders.

Our eyes have "cone" cells to distinguish color and "rod" cells which gather light. The owl's eyes contain mostly rod cells with a large amount of visual purple, a converter, which takes even the faintest glow and gives a chemical signal that quickly presents the owl with a slight impression. Human eyes get only an awareness of light.

Horned owls can see during the day, but as their great eyes are extremely sensitive to light, the iris nearly closing in harsh light, they prefer to move at night. They do, however, hunt during dark, cloudy days and sometimes during daylight hours when summer days lengthen. One late July afternoon at about three o'clock I saw a big horned owl suddenly appear out of the sky without a sound. Falling like a shadow, its feet outstretched, it struck a squirrel's nest in the top of a tall oak, then sailed away with the squirming animal in its talons.

The owl's body is covered with downy feathers that are so soft they seem to destroy sound. Noiseless flight is further aided with fuzzy flight feathers that deaden the whirring noises all other birds make. This well-endowed bird even has the additional benefit of longevity: One great horned owl lived in captivity for sixty-eight years.

Their talons are so powerful that they can easily drive through a person's hand, as New York State artist Wayne Trimm discovered. (Trimm also attested to the stamina of the horned owl after seeing one crash through the window of a moving bus, knock itself unconscious, then recover without any ill effects.) Naturalists Austing and Holt report that their assistant had a captive horned owl clutch his hand with such strength and tenacity that in order to break the hold they had to sever the owl's leg tendons.

In Minnesota's North Woods, Forest Service rangers who wore muskrat or coonskin hats were attacked by horned owls